Virus Hunter

C. J. Peters, M. D.
and Mark Olshaker

Virus
Hunter

Thirty Years of
Battling Hot Viruses
Around the World

ANCHOR BOOKS
A DIVISION OF RANDOM HOUSE, INC.
New York

First Anchor Books Edition, May 1998

Copyright © 1997 by C. J. Peters

All rights reserved under International and Pan-American Copyright Conventions. Published in the United States by Anchor Books, a division of Random House, Inc., New York, and simultaneously in Canada by Random House of Canada Limited, Toronto. Originally published in hardcover in the United States by Anchor Books in 1997.

Anchor Books and colophon are registered trademarks of Random House, Inc.

This book was written by Dr. C. J. Peters in his private capacity. No official support or endorsement by the Department of Health and Human Services, the U.S. Public Health Service, or the Centers for Disease Control and Prevention is intended or should be inferred.

The Library of Congress has cataloged the Anchor hardcover edition of this work as follows:

Peters, C. J.
Virus hunter: thirty years of battling hot viruses around the world /
C. J. Peters, with Mark Olshaker.
1. Peters, C. J.   2. Epidemiologists—United States—Biography.
3. Virologists—United States—Biography.   4. Hemorrhagic fever—
Epidemiology.   I. Olshaker, Mark, 1951–   .   II. Title.
RA649.5.P48A3   1997
610′.92—dc21
[B]   97-977
CIP

ISBN 0-385-48558-1
www.anchorbooks.com
Printed in the United States of America

10

To Susan and to my friends and colleagues,
all of whom have made my life so rich.

Contents

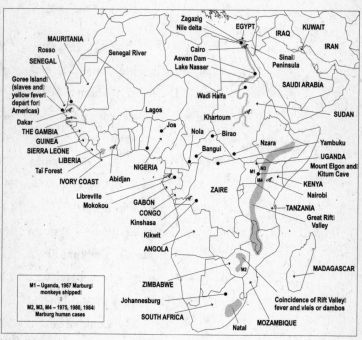

Zagazig
Nile delta
EGYPT
KUWAIT
IRAQ
IRAN
MAURITANIA
Rosso
Senegal River
Cairo
Sinai
Peninsula
SENEGAL
Aswan Dam
Lake Nasser
Goree Island
(slaves and
yellow fever
depart for
Americas)
SAUDI ARABIA
Wadi Halfa
Lagos
Dakar
Khartoum
SUDAN
THE GAMBIA
Jos
Nola
GUINEA
Birao
SIERRA LEONE
Bangui
Nzara
Yambuku
LIBERIA
UGANDA
Taï Forest
NIGERIA
Mount Elgon and
Kitum Cave
M3
M1
IVORY COAST
Abidjan
ZAIRE
M4
KENYA
Libreville
Nairobi
Mokokou
GABON
TANZANIA
CONGO
Great Rift
Valley
Kinshasa
Kikwit
ANGOLA
M2
MADAGASCAR
ZIMBABWE
Johannesburg
Coincidence of Rift Valley
fever and vleis or dambos
SOUTH AFRICA
MOZAMBIQUE
Natal

M1 – Uganda, 1967 Marburg
monkeys shipped

M2, M3, M4 – 1975, 1980, 1984
Marburg human cases

Kent Wagoner

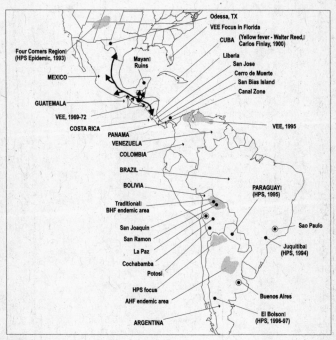

Odessa, TX

VEE Focus in Florida

CUBA (Yellow fever - Walter Reed,
Carlos Finlay, 1900)

Four Corners Region
(HPS Epidemic, 1993)

Mayan
Ruins

Liberia

San Jose

Cerro de Muerte

San Bias Island

Canal Zone

MEXICO

GUATEMALA

VEE, 1969-72

VEE, 1995

COSTA RICA

PANAMA

VENEZUELA

COLOMBIA

BRAZIL

BOLIVIA

PARAGUAY
(HPS, 1995)

Traditional
BHF endemic area

Sao Paulo

San Joaquin

San Ramon

La Paz

Juquitiba
(HPS, 1994)

Cochabamba

Potosi

HPS focus

AHF endemic area

Buenos Aires

El Bolson
(HPS, 1996-97)

ARGENTINA

Kent Wagoner

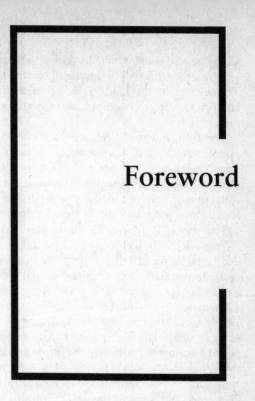

Foreword

FOR AT LEAST twenty years I have heard people say that the discipline I work in is a dying field and there's no career track—not very pleasant if that's the way you want to make your living, if you have put most of your professional life into that area, and if you can't think of anything you'd rather do. But there are a number of us working in the field of hemorrhagic fevers and other viruses in the natural world that infect man and we believe that it is a vitally important area for the future. In spite of our optimism (which may be the optimism of the brontosaurus) and deep belief in the need to continue, the number of gray heads around the conference tables is disproportionate.

At last, in the past few years we have had real champions. The Institute of Medicine studied the situation and declared that we needed to find the national political will to deal with emerging new infections, and many of the examples used in their report came from areas in which I have been personally involved. The Centers for Disease Control (CDC) initiatives to improve international surveillance and control of infectious diseases appear to be proceeding in

Congress and some money has been made available; it remains to be seen whether this will be sustained or amplified to the levels needed. The situation in the Third World promises to become worse, particularly in the area of overpopulation, and tax our resources even further.

Ironically, one of the greatest boosts has been the public success of popular books and journalistic accounts. Some of these books are very entertaining, but not really authoritative, and others are quite exaggerated. The common denominator is that they show the world some of our worries and also give a glimpse into what I have always found to be a deeply fascinating scientific arena.

This book is meant to speak particularly to those who ask, "What's it really like?" and "How did you come to do what you do?" Maybe we can dilute out some of those gray hairs around the table. It also tries to give some overview from my personal perspective as to what the real threats are from emerging viruses. I have seen a lot of sense and nonsense in the press and the Internet, but the simple fact is that no one has the real answers. There are, however, some straightforward things we can do to make ourselves more competitive against the viruses in the future and I have argued for some of these.

Originally I had intended to include more insights into the way we do science and some of the generalizations in this field, but the constraints of space, given the desire to also tell a story, were limiting. Some who participated in the events recounted here will have different ideas about what happened; these are my own perceptions, filtered through memory, and somewhat abbreviated to make them flow better for the reader. I don't wish you the waits during the incubation periods of the tests, or the long layovers in airports. Finally, a note about names: I have generally respected the medical rule of preserving anonymity or use of pseudonyms except where names have been used often in the press or are distant in time and space.

I am grateful to my new friend and the literary talent involved in this book, Mark Olshaker. I hope the many people whose lives were intertwined with mine understand that I had to choose some things that moved the narrative forward and not include some friends and times that will always be special memories for me. Above all, I trust that the readers realize how much we owe to the many scientists who preceded us, how many people were involved in the work cited in the book who were not mentioned by name, and the many other contributions of the people who may only be named here in passing.

From beginning to end, there was a core team involved with the creation of this book and great appreciation and admiration go out to all of them: our researcher and assistant, Ann Hennigan, who pulled it all together, kept us constantly on track, and whose work was extraordinary and far beyond the call; our editor, Roger Scholl, who saw what we wanted to do and continually encouraged us to achieve it; to Mark's intern, David Altschuler; to our agents, Cynthia Cannell and Jay Acton, who put the project (and Mark and I) together; and, of course, to my wife, Susan, and Mark's wife, Carolyn, for all their love, support, patience, tolerance, and good judgment. Carolyn has already been extolled in the forewords to Mark's many other books, but I want to say to Susan here that my gratitude for her contribution to the project and all the rest of my life for the past twenty years can't be expressed in writing, even with Mark's help.

And finally, to all of my friends and colleagues throughout my career, home and abroad, in universities, at NIH, at USAMRIID, at CDC, and in all of the brother and sister organizations around the world, I want to state my thanks, my respect, my pride to be among your ranks.

—C. J. Peters, M.D.
Atlanta, Georgia
December 1996

Life is short, the art long, opportunity fleeting, experience treacherous, judgment difficult. The physician must be ready, not only to do his duty himself, but also to secure the cooperation of the patient, of the attendants and of externals.

—HIPPOCRATES

He that will not apply new remedies must expect new evils; for time is the greatest innovator.

—SIR FRANCIS BACON

The pursuit of an idea is as exciting as the pursuit of a whale.

—HENRY NORRIS RUSSELL

Virus Hunter

Prologue

The Hoofbeats
of Zebras

THERE'S AN OLD ADAGE in medicine that goes something like this: *Common things occur commonly. Uncommon things don't. Therefore, when you hear hoofbeats, think horses, not zebras.*

Each of us has a worst nightmare, the kind of fevered dream that wakes us up in a cold sweat, overwhelmed by the sheer physical relief of realizing this dark and vivid reverie isn't real. Mine has to do with the zebras. What if the hoofbeats, this time, aren't horses? What happens if a deadly virus for which we have no treatment or cure explodes into the middle of a major city?

On Monday morning, November 27, 1989, it looked like my nightmare was coming to life.

I was in my office at USAMRIID—the United States Army Medical Research Institute of Infectious Diseases at Fort Detrick, Maryland—preparing for a meeting I didn't want to go to. As chief of the Disease Assessment Division, one of my many responsibilities was to distinguish between outbreaks of common, treatable bugs and rare, unknown diseases, between the horses and the zebras. Disease Assessment was a new division at RIID, housed in a temporary building, a prefab structure thrown up on a concrete slab. When any of us had to go to the bathroom, we had to go to other buildings, and until they caught us, we used to empty our coffeepots out the fire escape. By government logic, if our temporary facility had had plumbing, it would have been a permanent building and Congress on high would have had to approve it.

My colleague Peter Jahrling and I had different philosophies about difficult meetings. I put them off as long as I could, but once I schedule them, I go. Peter, the division's senior research scientist, would schedule anything in a heartbeat but would then try to find some excuse at the last minute not to go. This was one neither one of us wanted to go to: a meeting in Rockville on the simian immunodeficiency virus (or SIV) project with Don Burke and the Walter Reed Army Institute of Research people. They were top-notch scientists and had the same goals we did, but we had disagreements with them about the science, the funding, and the politics. But the bottom line was that simian immunodeficiency was a good model for studying HIV, and so it was an important project. If we could get some insight into the mechanism of disease in monkeys, it could help us figure out how to treat it in humans, and this could also be a real breakthrough for vaccine development. Peter and I had made a deal with each other that no matter what else came up, we'd both go to all the meetings.

About ten minutes before we had to leave for the half-hour drive to the suburb north of Washington, D.C., I was rummaging around my office checking to make sure my uniform had all its costume jewelry in the right places. I've never been a very snappy dresser, maybe because it was so hot in West Texas, where I grew up. I've always favored jeans and bright-colored shirts with strong prints rather than a regular uniform—a habit not always appreciated by my superiors in the military hierarchy. But I had picked up a tenacious form of athlete's foot fungus years before, working in the jungles of Central America, which I used as my excuse for wearing casual clothes and sandals with white socks to the office. I didn't start wearing a uniform regularly until I became a colonel, at which point I had to start setting a "good example" for the others. I still wasn't very good at it, though, and every time I had to represent RIID to the outside world, I had to double-check to make sure everything was where it belonged.

When Peter showed up at my door, I thought he'd come with some excuse why he couldn't go. With him was Tom Geisbert, a twenty-seven-year-old grad student Peter had just recruited away from John

White in pathology to give us electron microscopy capability. John had worked with Tom's father, an engineer, "behind the fence" in the old days when Detrick housed the army's secret biological warfare program.

Tom was holding a manila folder. "C.J.," Peter said, "you need to take a look at this."

Tom handed me the folder. There were several eight-by-ten glossy photos inside. Electron micrographs. As I flipped through them, inner alarm bells were beginning to sound.

Peter gave me a minute, then asked, "What does this look like to you?"

I saw rod-like particles in the cells, some curved at the end like a shepherd's crook, others in bizarrely twisted shapes. "It looks like a filovirus," I responded apprehensively. I was suddenly gripped by what we refer to as "the pucker factor," an uncomfortable and unpleasant tightening of certain sphincter musculature. "Where does it come from?"

"These are the samples from Reston, Virginia."

Oh shit, I thought. I knew that Dan Dalgard, the supervising veterinarian for Hazelton Research Products over there, had contacted Peter after some of his monkeys started dying at their primate quarantine unit. Reston was an upper-middle-class bedroom community, one of the first of the "new towns," about twenty miles outside Washington, D.C.

Hazelton imported macaques from the Philippines for medical research throughout the United States. It isn't unusual for a few monkeys from a shipment to die during the quarantine period before they can be certified disease-free and sent out to the labs. But normally, when they die it's either something so vague that you never find out what the cause was, or they have weight loss and diarrhea before they die of dysentery. These monkeys were bleeding internally and externally before dying a very unpleasant, dramatic death. They had firm, enlarged spleens, which probably meant they were packed full of white blood cells, a healthy immune response. Dissecting the organs had been like cutting through sausage. Equally troubling to Dan Dalgard, all the bleeding deaths had occurred in one room—Room F—of the single-story, brick-faced quarantine facility. He was worried he was dealing with an outbreak of simian hemorrhagic fever (SHF), a disease which doesn't affect humans but is absolutely devastating to certain species of monkeys. He had sent samples from all the

monkeys that had died in Room F over to Peter to work up, looking
for SHF virus. And they'd found some.

Electron microscopy requires hours of careful processing of tissue
or cells by a skilled technician before the test can be carried out.
Looking at the slides under the electron microscope, Tom had discovered something else disturbing the cells, and it sure as hell looked like
a filovirus. There were only two known filoviruses—Marburg and,
even worse news, Ebola, two of the deadliest of the African hemorrhagic fevers. Both caused the kind of symptoms the monkeys in
Room F were dying from. Only they caused it in human beings.

The first documented outbreak of a filovirus had occurred at a
vaccine factory in Marburg, Germany, during the summer of 1967.
Three employees thought they'd come down with the flu. But by the
next day they'd developed severe diarrhea and spreading rashes.
Then they started vomiting blood.

And that wasn't as bad as Marburg's sister virus: Ebola. During
the summer of 1976, 318 people came down with Ebola in a remote
area of Zaire. Two hundred and eighty of them died horribly, including nurses and Belgian missionary nuns who tended the sick in the
remote village hospitals. Ebola produces one of the worst deaths I
can imagine.

Tom's discovery had suddenly become a major pucker factor.

"How do you know this isn't just a lab contaminant?" I pressed
Peter.

"There's no other Marburg or Ebola in the lab," he said.

Oh shit, I thought again.

There are four laboratory levels of biological containment, each
one keyed to the potential threat to human life. The lowest Biosafety
Level, BSL 1, is for organisms or agents not known to cause disease.
BSL 2 is for viruses and bacteria which cause only mild disease or are
difficult to contract in a laboratory setting, such as diseases which
don't transmit by aerosols. In BSL 3 the bets are higher. It's constructed as a "box within a box," featuring constant one-way airflow
to keep the germs inside, and laminar flow hooded work stations to
protect gowned and gloved researchers against organisms which
cause serious or fatal disease, such as rabies or Rocky Mountain
spotted fever, but for which vaccines or treatment is available. BSL 4,
the maximum containment lab, is the hottest of the hot zones. Here
researchers who've undergone a battery of inoculations move about
in full-body "space suits" connected to a constant air supply and
wear double gloves taped to their sleeves as they examine and work

with life-threatening biological agents like Ebola virus for which there is no vaccine, no magic bullet drug, and no cure. Decontamination tanks and filters are designed to kill off any biohazard waste from the lab before it reaches the outside world. Work in the BSL 4 lab is gut-checking, high-stakes science, where any oversight or slip-up can have disastrous consequences. It takes a particular personality, going far beyond scientific ability, to work regularly in maximum containment. There are only two Level 4 labs in the United States— one at the Centers for Disease Control in Atlanta, the other at USAMRIID.

"Have you been working up any other human samples from Africa?" I asked

"No," said Peter.

"Any other unknown viruses or mystery samples?"

"No."

The only other viral agent in the Level 3 lab was Venezuelan equine encephalitis, and never in a million years would you mistake that for a filovirus.

I turned back to Tom. "Talk to John White. He's been looking at filoviruses for years now—show him what you've got and see what he thinks. I think you're rock solid at what you do. But I don't think you have John's experience, and we're going to use every resource we have to be sure."

There had never been a filovirus outbreak in the United States. Figuring out if we had one now was going to take several hours. In the meantime, there was nothing more for Peter or me to do. So, reluctantly, we dragged ourselves out to the meeting in Rockville and spent the afternoon hacking our way through the logistic tangles and conflicting opinions over protocols for SIV monkey experiments that might get started in a few weeks and would last months before we got any answers. But every word that we said was now overshadowed by what was going on in Reston. We were puckering big time, though we kept it to ourselves. *Don't let this be what I'm afraid it is.*

The moment we got back to Detrick, I called Tom Geisbert.

"What did John White say?" I asked urgently. "Did he think it was a filovirus?"

"Yes," Tom replied. "He said it looked like Ebola."

It was as if my heart stopped. We had Ebola, uncontained. *Right here in River City.* My mind started racing: We've never seen this virus in monkeys before. How did this human-killing African virus get to a suburb nestled against the heart of the capital of the United

States? How many veterinarians had treated these monkeys? How many caretakers had come in contact with them? Where in Virginia, Maryland, or Washington, D.C., had they sent all the previous lab samples and how many technicians had worked on them? Ebola spread by close personal contact in a ripple effect, an exponential network. We could find out who the monkey caretakers were. We could find out who their families and friends were that they might have exposed. But how in hell were we going to find the contacts of the contacts?

Gene Johnson, a research scientist at RIID, and I had been playing around with Ebola in the lab and had a profound respect for its power. We'd shown that under laboratory conditions we could generate an aerosol that infected primates. We had seen the virus spread from one monkey cage to another with no direct contact in the middle of a BSL 4 laboratory. If what we saw in the lab happened in the outside world, we were in serious shit. In point of fact, there was no evidence that this could or would happen in the real world. But we also knew we'd all been surprised more times than we cared to add up. And most of the time, the surprises were not happy ones.

"Did you tell John where it came from?" I asked Tom.

"No."

"Good. I want this moved to Level 4 immediately."

There was no need for any of the three of us to say anything out loud, but we were all thinking the same thing. *Could this finally be the big one we've been dreading our entire careers? Ebola killed 90 percent of the folks who came down with it in Zaire in 1976—people who hideously bled to death while their internal organs shut off one by one—and we still don't know where it comes from or what the animal reservoir is. All we know is that once this thing climbs aboard an airplane, it could be anywhere in the world within twenty-four hours. And for now, we've got a fucking killer virus loose just outside Washington, D.C.!*

We began hearing the approaching hoofbeats of zebras.

1

The Killer
Without
a Name

THE RESTON EPIDEMIC was neither the first nor the last viral epidemic to raise my pucker factor and send alarm bells ringing throughout the medical community. Only three years later there was another one. A few things were different. I was no longer at USAMRIID; instead, I was chief of the Viral Special Pathogens Branch at the Centers for Disease Control in Atlanta. And this one didn't start with dying monkeys. This time, humans were the first to die.

Let me try to put this in some context for you.

One day your child, your parent, your spouse, or your lover—the person you cherish most in the world—is vigorous and healthy and full of life. Then he or she comes down with a headache, some fever and body aches, his or her chest feels heavy, breathing becomes labored. They complain of vague symptoms that get worse and worse. Sometime later, they collapse.

Twenty-four hours later, they're dead.

That's exactly what happened around the middle of May 1993 in the high plains of the American Southwest, and I and others at CDC were charged with finding out why—and fast.

The first of two index cases—square one for the outbreak—was Florena Woody, a twenty-one-year-old Navajo woman in superb physical condition, a long-distance runner who managed the Santa Fe, New Mexico, Indian School track team. She collapsed suddenly with what we call ARDS—adult respiratory distress syndrome. ARDS is a catchall diagnosis we can use even when we don't really know what the problem is. Within hours after her collapse, she had died.

Five days later, her nineteen-year-old fiancé, Merrill Bahe—himself a nationally known distance runner, whom she had met and fallen in love with at the Santa Fe school—collapsed on the way to her funeral from their home in Littlewater, New Mexico, and died before reaching the Indian Health Service facility in Gallup, leaving behind their five-month-old son, Maurice. It's the kind of thing that scares the hell out of any good doc, and Bruce Tempest, chief of internal medicine at IHS-Gallup, was a very good doc.

It would have been an easy thing to miss. One case of something odd doesn't usually ring any bells in the medical community. Two are still pretty easy to shrug off, although they do give you a clinical course to make comparisons; there were also some interesting clinical features. But this was just the beginning. Within a week, Florena Woody's brother and his girlfriend developed similar symptoms. Her brother hovered near death for days but somehow pulled through: his girlfriend almost died. Florena and Merrill's baby also developed fever, but later proved to have only a childhood respiratory infection.

What made one person die but allowed another to fight off the disease and recover? It was a question that only compounded the mystery and the fear. *If I suddenly start having trouble breathing, am I coming down with the disease?* residents in the area wondered. *Am I going to die?*

It was when Merrill was brought in to the Gallup clinic that Dr. Tempest first realized the illness was different from anything he'd ever seen. So he started networking, talking to other physicians, which was how he found out about Florena so quickly. And when he did, there was a sudden recognition among his colleagues that otherwise healthy Navajo Indians around the Four Corners area—where Utah, Colorado, Arizona, and New Mexico come together—were dying suddenly of respiratory failure. Tempest immediately notified the New Mexico Department of Health, the Office of the Medical

Examiner, and the IHS epidemiologist—thirty-five-year-old Dr. James Cheek, a Cherokee who had recently arrived in New Mexico after spending two years as an Epidemic Intelligence Service, or EIS, officer with CDC in Atlanta.

The Office of the Medical Investigator already knew about the problem. I'm originally from Texas, and I returned there after med school for internship and residency training. Texas, like most states, has few laws governing the qualifications of the official who sees to the dead. In many places, the county coroner could be the guy who owns the drugstore and who simply stood for election. He may be highly qualified to run a drugstore, but not necessarily to conduct a postmortem examination—or even know when one is warranted. But one of the fortunate things about dealing with New Mexico is the state law that established the Office of the Medical Investigator, with far-reaching authority and discretion to investigate. Richard Malone was an investigator for the DA's office in Gallup, but he was assigned to be the local eyes and ears for the pathologists. When he looked into Bahe's death and learned that his wife had died with a very similar syndrome, he was startled. It was about 1:30 and he knew that Florena was to be buried at 2:30. After a quick contact with the pathologist, Dr. Patricia McFeeley, he went to the family and carefully explained the situation to them. Their loved one had died on the Navajo reservation, so Malone had no jurisdiction, and the sensitivities surrounding the dead in their culture were extremely against the idea of disturbing the body. In the face of all this, they made the admirable decision to permit an autopsy. McFeeley did both posts that evening and was struck with the similarity and even recalled a similar case from April. The lungs were so edematous—so filled with fluid—they weighed twice what they should have.

There was a cluster of unexplained deaths or near-deaths with similar symptoms in the area: headaches and sudden fever leading to fulminant pulmonary disease, causing both lungs to give out, requiring extra oxygen and then an intratracheal tube to aid breathing. In many of the cases, even that didn't do much good and the gruesome end came quickly as the patient drowned in his own plasma. Now the alarm bells were ringing all over the Southwest.

The forces of public health began to marshal in a textbook example of cooperation. The Navajo Nation, Indian Health Service, and New Mexico Department of Health went to the home of the dead runners. What clues could have been left to speak for them?

When none of the cases fit a clear-cut clinical pattern for an infection, one thing that immediately came to mind was the possibility that some sort of toxin was at work. Spanish oil syndrome, for instance, had become famous in epidemiology circles for causing pulmonary disease, fever, and more than 600 deaths in Spain between 1981 and 1983. It was ultimately traced to adulterated oil that had been sold for cooking; the contaminant that made more than 20,000 people sick is still debated.

Cheek had already done some investigation in this area, and in his thoughts two likely culprits for the Four Corners outbreak were phosgene and phosphene, chemical compounds which were known to produce ARDS symptoms in healthy people within a relatively few hours after exposure. Phosgene gas was used by the Germans in World War I and its horrifying effects contributed to the international outlawing of chemical warfare. But it had also been banned in the United States, and getting enough of it together to cause deaths around Four Corners would have been difficult and unlikely. Phosphene, on the other hand, Cheek discovered, was legal, and was used to poison prairie dogs, which were considered a nuisance by many and could be carriers of plague. It seemed a logical answer. A toxin was a likely culprit and phosphene seemed the strongest possibility.

But when Cheek visited Merrill Bahe's trailer home in Littlewater with the team, he discovered no evidence of phosphene containers, dispensing equipment, or anything else that might be related to phosphene. Furthermore, the trailer was in appalling condition. Since Bahe's death, it had been overrun by rats or mice, whose droppings littered the place.

Using no special precautions or even protective gloves, they took samples from around the home, including food, dishes, and the rodent feces they saw on the floor. It was a rash move, but it was weeks before they realized it.

When toxin screens didn't pan out and the state people couldn't come up with anything substantive to explain the outbreak, Cheek called one of his CDC colleagues, epidemiologist Dr. Rob Breimen. That's often the way CDC gets involved. The world of epidemiology and the public health community is a relatively small and tight one and many people in it know one another well. The state health departments are well represented by folks who have rotated through the two-year EIS program, where they're trained as medical detectives, fanning out around the United States and overseas investigating medically mysterious deaths and disease outbreaks.

As I read the dry clinical summaries and the sketchy outlines of the autopsy reports we'd received by fax, I could imagine the disease process. These patients were very sick during the early phase, much more so than the clinical description we later put together would indicate: "fever, myalgia, malaise . . ." Their "myalgia" must have been the sort of deep muscle aches that in the last century had led to dengue fever being dubbed "breakbone fever." "Malaise" just means "feeling bad," but that doesn't bring some of these tough Westerners to the clinic during the prodrome; they must have felt *really* bad and not just some droopy feeling or malaise of the spirit.

Then, when the actual lung phase of the illness began, it would have been much worse. Their airy, almost fluffy lung tissue would have become increasingly heavy and demanded more and more effort for every breath as the fluid from their blood seeped into one of the most critical organs of the body. This vital function is "alarmed," just like an expensive Porsche parked on a New York City street, and when the work of breathing increases, the signals to the brain are frightening. Essentially, they are telling the person to get a deeper breath *now*, and the process was on a down-hill course. As the tissues of the lungs swelled with plasma from the blood, it seeped out into the tiny air sacs as a liquid, making it harder and harder to get the oxygen the patient needed for his very life from those air sacs across into the blood. Quarts of fluid accumulated in the chest cavity around the lungs. A normal lung is whitish, actually a delicate pink, and it cuts like an extremely soft sponge. But pathologists described the cut surface of these lungs as almost like meat, and the color was dark bluish-red from lack of oxygen. The fluid exuded from the lungs into a straw-colored puddle as they lay on the cold stainless steel of the necropsy table.

The state health department and the University of New Mexico had done extensive lab studies on the deaths which largely ruled out all the usual suspects. None of the experts could think of anything that matched up with the clinical details of the mystery illness: not influenza, not anthrax or any other bacteria, none of the normal viruses. So what the hell do you do?

Well, one of the things a state agency can do is to ship samples to us and ask that we do some tests uniquely available at CDC: specialized flu tests, mycoplasma tests, even bubonic plague confirmation. We see all the weird stuff; maybe we can figure out what's going on. The main thing is, people are dying; they're terrified. Someone has to figure this out.

Plague is a flea-to-rat-borne bacterium we now know as *Yersinia pestis*. It devastated Europe in the fourteenth century, killing off, by the best estimates, about a third of the population. *Yersinia pestis* was brought to Europe by the Mongol invaders, but it's a restless bug and the black rat, and the Norway rat that it pairs with in the cities, are invasive creatures. By the 1500s there were suspected plague epidemics in Spanish settlements in Mexico, and by the 1600s the Pilgrims had the "advantage" of something resembling plague to clear the way for their settlements. Today, plague has set up its own cycle among the chipmunks and prairie dogs in the Southwest, and there are about twenty cases among people every year in the United States, most of them in rural New Mexico. Could this be pneumonic plague? Well, no one knew, but the New Mexico Health Department was probably as good at finding it as any in the world. Maybe they had missed something, though. They were open and professional enough to let someone else take a shot.

CDC is actually eight individual "centers" that deal with public health concerns ranging from chronic disease to environmental health, to injury prevention, to occupational safety. My branch, Special Pathogens, is in the Division of Viral and Rickettsial Diseases under the National Center for Infectious Diseases, or NCID.

On Friday, May 28, Ruth Berkelman, deputy director of NCID, called a meeting of representatives from all the center's divisions, including several of the other Viral and Rickettsial branches. She even roped in some folks from the Center for Environmental Health because of the toxin issue. The mission was to review the data which had come in piecemeal from various sources, including the health departments of New Mexico, Arizona, Colorado, and Utah and the Indian Health Service. There were about twenty of us at the meeting, in the personal office and conference room of the head of the Division of Bacterial Diseases, on the fourth floor of Building 1, the main CDC building, which looks out over Clifton Road. It was a nice room, lots of windows letting in the bright May sunshine, several tables circled around a desk—sort of standard government-issue furniture but better maintained and cared for than what most of us have. I felt like a mole coming up from my cluttered, windowless office in the basement of the lab building.

Individually and collectively, we are a mix of medical detective, scientist, healer, and just plain human being, so it's not unexpected that there's a kind of schizophrenia operating at these meetings. On one level, you know that lives are at stake and it's urgent to uncover the identity of the unknown killer. On another level, though, you know the only chance of solving the mystery is through good, methodical science. You've almost got to put blinders on, to remove yourself one step from the human anguish. You've got to feel for the victims, but not be crippled by your emotions. You've got to go systematically through the clues. Like homicide detectives, medical detectives often start from the same place: what the body itself tells us.

ARDS is usually a reaction of the lung found in severe burn or accident victims, people with overwhelming bacterial infections of the bloodstream, and is most common in the elderly. It's not the kind of thing you expect to see in young, healthy men and women, much less athletes who've extended the endurance of their hearts and lungs through years of rigorous training. Adding to the sense of urgency, local and national news organizations, such as the *Albuquerque Journal* and CNN, were already releasing stories on what they called the "Navajo disease" or "Navajo flu."

Flu can be another of those incorrectly used catchall terms when no one knows quite what's going on. It may sound benign to uninformed ears until you remember that Legionnaires' disease was first reported as "Legionnaires' flu" and AIDS was first labeled "gay flu." Those of us in the business always think of the specific virus, influenza A, and keep in mind that while this flu is usually not life-threatening in and of itself, the worldwide influenza outbreak of 1918 killed more people in a year than any other event of the twentieth century—war, natural disaster, famine, or epidemic.

There was another factor at play here, equally disquieting. Some of us, particularly myself, were used to jumping on airplanes at a moment's notice and flying off to Africa or South America or some other remote region at the first report of a mysterious and deadly disease outbreak. But what was particularly terrifying was that now it was happening right here in our own country. People were dying and we had no idea why.

A number of people in the room had firsthand recollection of the Legionnaires' disease outbreak of 1976 that had everyone stymied for months. *Legionella pneumophila,* as it came to be called, was now known to be a common respiratory infection and occasionally the cause of the adult respiratory distress syndrome that had struck hundreds of men and women who had gathered in July 1976 in Philadelphia for an American Legion convention at the stately Bellevue-Stratford Hotel. Many of them came down suddenly with acute pneumonia. Despite all the standard treatments and the best medical support, thirty-four of them died. It took months for two of the best research scientists in the world, Joe McDade and Charles "Shep" Shepard, to figure out that what they were dealing with was a particularly elusive bacterium, one which had stymied all the conventional efforts to get it to grow in the lab. And when they'd finally identified it, other medical mysteries were almost instantly solved. By matching up frozen samples kept from two previous cases, it explained the deaths of people at St. Elizabeth's Hospital in Washington, D.C., in 1965 and, three years later, in a Pontiac, Michigan, health department office.

But the Legionnaires' victims were mostly older people with a host of risk factors: they smoked, they drank, they had chronic lung problems or immune systems battered by age or disease. The Four Corners victims were mostly young and healthy. If it took as long to crack this case as it had Legionnaires', we would all be in for a rough time. As it was, with news of the mystery disease spreading, the Gallup Indian Medical Center was seeing about 2,800 patients a day—up more than 2,000 from a normal day!

Altogether, we had data from maybe a dozen patients, including the two index cases. At the meeting we passed around summaries of the clinical findings and autopsy analyses. We looked at dates of onset of symptoms and asked what these people had done or hadn't done, where they were, anything that might distinguish them from the people who didn't get sick.

Basically, the patients were predominantly Native American, all rural residents, who had a very rapidly progressing disease that led their lungs to fill up with fluid, preventing air exchange. They died, often agonizingly, literally gasping for breath. If you looked at the X-rays, what you saw was a rather terrifying condition known in medical slang as "whiteout." On a normal chest film, you see the black lung field, the white ribs, and grayish lung markings. But when the lungs are full of liquid, that increases the radiological density, so that when you shoot the X-ray, it looks like a snowstorm. What you're seeing are lungs that can't breathe any longer. You're seeing the insides of a person who's dying.

The infectious disease people tended to favor a toxin. The toxin people said it had to be an infectious disease. Within the infectious disease group, the bacterial people leaned toward a virus and we viral people thought it sounded more like a bacterium. The influenza people favored a mycoplasma, but the mycoplasma people weren't so sure about that; maybe it was a chlamydia. It's a sobering feeling when you've got all these experts gathered around a table, people who bring tremendous knowledge and expertise to specialized, defined areas, and essentially they all shrug, throw up their hands, and say, "Not mine."

In an individual patient, we couldn't rule out a fulminant influenza A, either the virus itself or complicated by secondary bacterial pneumonia spreading over the lungs, similar in some ways to the 1918 variety. The mortality rate can be high. But if it was influenza, there should be people coughing all over New Mexico. You should be able to hear the coughing down below from an airplane. We weren't getting those reports. In addition, Steve Young at the University of New Mexico, as well as the state health department, had cultured for influenza, and CDC had looked as well, using very sensitive techniques. And no one had found any evidence thus far.

"What did the victims have in common?" we asked ourselves. Most were Native American, most were young and previously healthy, and almost all were rural. Were any of these factors significant?

The Native American issue had already become a publicity hot button. As the news spread on the so-called Navajo disease, Navajo men and women were perceived as carriers of death. We received reports of non-Navajos refusing to enter Navajo-owned businesses, of non-Navajos refusing to serve Navajos or wearing rubber gloves or masks in dealing with them and generally treating them like lepers in the Middle Ages. A private school in Los Angeles canceled a visit to their students that had been planned for twenty-seven Navajo pen pals.

At the same time, paranoia was spreading within Native American communities—with rumors that this disease was a biological killer agent developed by the military to wipe out undesirable populations. Stories had survived for more than a hundred years of the U.S. cavalry giving Indians smallpox-contaminated blankets. While I understood the historical basis for the fear, I also knew how quickly fear and distrust could poison the atmosphere and complicate our job of locating the disease source.

My own feeling from the start was that the fact that most of the victims were Native American was incidental. The Navajos predominated in the region, particularly in the rural areas, so naturally they were the ones most affected. What seemed more significant, I felt, was that the large majority of victims lived in a rural environment.

That tends to suggest some kind of rural or agricultural occupation that could be a specific risk factor or, just as likely, a zoonotic reservoir—that is, a rat or mouse or some other animal carrying the disease and spreading it to humans.

Life was very traditional on the Navajo lands. They call themselves *Dine' Bahane'*, or "the People," and their past is interwoven with the land. They have lived a pastoral and agricultural life for so many generations that their gods have local "addresses" at nearby mountains. Many Navajo prefer to live in scattered, isolated camps with only their family or extended family, and this, of course, complicates the delivery of modern services such as telephone lines, electricity, piped water, and medical care. This culture is under pressure from the intrusions of the surrounding society and the growth of the Navajo population which threatens to overwhelm the carrying capacity of the ecosystem, as well as from such traditional urban threats as alcohol and drugs. The isolation in their lives also brings a vulnerability to microbes and the Navajos have their share, particularly tuberculosis, a disease brought from Europe and to which Native Americans are probably inordinately genetically sensitive.

As we kicked the disease possibilities around the table, I said, "Well, you know, I don't think this is one of our Special Pathogens Branch's diseases. But if it was, I would have said before the meeting that it was probably due to an arenavirus, because arenaviruses are usually rural and rodent-borne."

Arenaviruses, from the Latin word for sand, are so called because the virus particle is round and under the electron microscope appears to contain grains of sand. We classify several of the diseases they cause as hemorrhagic fevers, like about a dozen other infections from members of different viral groups such as Ebola and Marburg, two devastating filoviruses. In its most critical form, hemorrhagic fever can be as dramatic and relentless as anything you will ever see in medicine.

The impact of hemorrhagic fever on the body is swift and severe. It comes on abruptly and leads you on a downward slope as you feel worse and worse, with scattered symptoms from the many vital organs involved. The sense of fatigue is numbing, as though you were crushed under a boulder. Fever saps your will to work or play. The skin is flushed and so sensitive that you don't want a bedsheet to touch it. When you try to walk you feel dizzy and your balance is uncertain. Fortunately, you don't know a lot of the specifics that are going on inside you: your liver function tests will reflect small areas where the tissue is rotting, urinalysis shows microscopic evidence of countless foci of bleeding within the kidneys, the surfaces of the organs are covered with pinpoint hemorrhages and oozing plasma. These small bleeding points are one of the key features of hemorrhagic fevers—they are the visual evidence of the innumerable sites of damage to small blood vessels throughout the body and they are also present on the skin and the surfaces of the eyes. This diffuse destruction is the true hallmark of the hemorrhagic fever syndrome and is what leads to the failure of many organs in the body and, ultimately, to death. The dramatic hemorrhages seen in many patients are just a reflection of the end stage of the underlying process of multifocal microvascular damage, a death of a thousand cuts, if you will.

The fragile blood vessels around the gums and on the soft mucosal surfaces of the gastrointestinal tract begin to break with the slightest provocation, resulting in an ooze of blood. Sometimes this breakdown goes further and the ooze becomes a torrent, with the patient vomiting blood and passing blood in bowel movements. There is often dramatic bleeding under the conjunctivae, making the eyeball itself appear red and bloody.

The classic hemorrhagic fever is yellow fever, and it was so well known for gastric bleeding and vomiting blood, as well as the raw fear and denial it engendered in many, that Dr. Ashbel Smith, the surgeon general of the Texan army, described the 1839 Galveston outbreak as follows: "The efforts of the citizens . . . were paralyzed by the absurd denial of a few who feared their pecuniary interests would be damaged by a knowledge of the existence of yellow fever among us, aided by the gross ignorance of others, who in their pointless hostility to the name of yellow fever, declared the recent epidemic to be the Plague. They were, however, most signally rebuked by the disease stamping almost every fatal case with its unequivocal seal of black vomit."

This is the power that hemorrhagic fevers have had over the general population and the medical community ever since they were first recognized.

These many tiny foci of damage make it virtually impossible to treat some of the potentially lethal effects of the disease. As the patient slips into shock from the leakage of fluid and loss of blood into tissues (and sometimes externally), the physician can transfuse and try to push IV fluids, but the damaged vessels just leak the fluids back out into the tissues and sometimes into the lungs, only complicating the problem. The heart is also affected, and so even the passing improvement in blood pressure bought at the expense of expanding the vascular system is not really going to help in getting the blood to the many parts of the body where its oxygen and nutrients are needed.

You can think of the heart and blood vessels as being like a farmer in a drought. He would like to irrigate everything with life-giving water, but as the well dries up he progressively shrinks the circle of life to the most valuable crops. In the case of the body, its most protected sites are the brain, heart, and kidneys. So as the perfusion to the skin disappears, it blanches, becoming cold and clammy. The liver and intestines don't get the blood they need and their function decreases. When the kidneys are starved and urine formation stops, you know that the last protected areas are in danger. By this time the patient is delirious or comatose and the physician is not worried about his psyche but rather restraining his thrashing around, hoping there's a miracle around somewhere close. And sometimes there is. These patients can be very, very sick, literally at death's door, and pull through. One of the truly gratifying things about the fight to save their lives is that if they recover, they almost always recover completely. But in too many cases, they die.

The Four Corners disease wasn't pretty either. But hemorrhage wasn't a major part of the clinical description of the mystery. Yet I wasn't willing to let arenaviruses go completely. After all, I had cut my teeth on them in Panama as a young physician and you don't easily part with such intimate acquaintances. I'm a sentimental person and for me they were like an old girlfriend: there might not be much of a future, but I was never going to give up the memories.

"We don't know how many arenaviruses there are in this country," I admitted. "It's never been carefully looked at. Outside North America, some of them can cause respiratory collapse as part of the clinical picture, and some arenaviruses like Lassa fever can kill you without causing a lot of hemorrhage."

Some of the laboratory findings from the field were consistent with an arenavirus infection. The patients had low blood platelet counts and shock, both of which fit the pattern. But what spoke against these particular viruses was the tremendously large number of white cells—part of the body's defense mechanism—circulating in the blood, in some cases as high as 50,000. Medical dogma holds that bacterial infections cause high white counts but that viral diseases either don't change white blood cell counts at all or else they lower them. This dogma is incorrect, but arenas still didn't fit.

I did know of one type of virus that regularly caused high white counts: hantaviruses. Hantaviruses were known to cause human disease throughout Asia and in many areas of Europe. Some of the outbreaks had a 5 to 10 percent kill rate: bad, but not nearly as bad as what we were seeing in the Southwest. There was another problem with this theory. All the known hantaviruses attacked the kidneys, not the respiratory system, and they were always associated with hemorrhage. Typically, the disease would start out with some fever and severe body aches. Then when the immune response attacked the virus, the "by-products" of the struggle would lead to diffuse damage to the capillaries—the small blood vessels—and severe shock, followed by kidney shutdown. Patients could die from the bleeding that followed the capillary damage, shock, or renal failure, but usually pulmonary disease didn't dominate the picture as it did in our patients. It didn't add up.

"This probably isn't anything of ours," I told Jay Butler, an epidemiologist with the bacterial branch. "It's probably a toxin. But if it is one of our viruses, I would have to bet on a hantavirus."

What it really looked like to me, I told Jay, was a bacterial disease that triggered what's called a superantigen reaction in the lungs. One example was staphylococcal enterotoxins, which we'd worked with in the army in aerosolized form and which can really knock you on your ass with rapid-onset shortness of breath and fever. It was easy to see such a reaction causing other effects, including shock. As a matter of fact, this type of mechanism is suspected as the cause of toxic shock syndrome, identified initially with using highly absorbent tampons.

But we also had to face the prospect that we could be dealing with a new, previously unknown disease. If that was the case, all bets would be off.

That's when your epinephrine level starts building. I went into that meeting with the expectation that this was an influenza epidemic and they'd bagged three or four people with a rare but overwhelming syndrome. "Big deal," we'd all say to ourselves, "not my problem." In my own case, I'd add, "I'll go back downstairs and try to figure out how to get enough travel money for our Lassa fever project in Sierra Leone."

But what if the hoofbeats we were hearing weren't from the proverbial horses? What if it was zebras after all? Whether you're in the military or public health, there's always that mixture of fear and exhilaration when something big might be happening. You don't want anything bad to happen, but if it's big, you don't want to miss it and let someone else get the action. "Dear God," we all secretly implore, "please don't let this happen. But if it does, let it be on my watch."

Collectively, we quickly decided we needed to conduct a full-court press. We couldn't afford to bet wrong. Maybe the epidemic curve would quickly fall off. But maybe it wouldn't. What if it climbed into the hundreds, or thousands, and spread exponentially? You can't let something like this run unchecked and unchallenged and just assume that everything is going to be okay. Those who assumed in the early 1980s that AIDS would go away assumed dead wrong, and we've been paying the price going on two decades now.

The more we heard, the more concerned we became—and the more certain we needed to get involved. But the politics can be delicate. Unlike the Internal Revenue Service, we can't come unless we're asked. And local health agencies, much like foreign governments, don't always want CDC intruding into their investigations. There are good reasons for this, including the plain fact that it is their responsibility. They often have excellent staffing and they always have a better appreciation for the local circumstances. If they can handle a situation expeditiously, then they shouldn't bother the feds and they deserve all the credit.

Ruth Berkelman, the chair of the meeting, whose specialty is emerging diseases and who is a tough and resourceful epidemiologist, took the bull by the horns.

"Most of the deaths have been Native Americans," she pointed out, "and they were on a federal reservation, which gives us jurisdiction. I have already had a call from the Indian Health Service and they are alarmed at the possibilities, so while we wait for all the loops to be closed, we have to get started."

Hours later the states themselves called up with IHS and essentially said, "We've done everything we can and we'd like for you to come help us."

There followed, as there does in any of these situations, a period of intense horse trading at CDC. With all the cutbacks, as well as the shift in resources to AIDS research and prevention, the rest of infectious disease had taken some major hits both in staff and in budget which left us all stretched to the limit handling everyday responsibilities. This was going to be big. Like the American military in the Gulf War, one of the things I had learned at CDC is that you don't go into a major, fast-moving, highly visible investigation halfway. You move quickly and provide enough resources to get on top of the situation immediately. There is a real place for small, targeted programs that pay off handsomely over time for small investments, but it's not in a situation like this.

CDC was not going to let people die just because the budget was tight. Although there was no formal emergency fund in NCID, we are essentially given a credit card with no spending limit when there's an emergency facing the American public. However, there is no way instantly to buy trained, experienced people with that card—as I was to learn in spades. Somebody's got to ante up people to go to New Mexico. If the bacterial branch goes, it means that the work on Legionnaires' disease and drug-resistant pneumococci stops. If the Special Pathogens people go, then we stop working on isolating strains of Ebola or research into a vaccine for Lassa fever, a huge problem in West Africa, which we hoped could be a prototype for other vaccines. These days, medical resources are far from limitless. So whatever we decide to do, the impact on other parts of the nation's and the world's health is not trivial.

Everybody wants to help, but, selfishly, you start thinking about your own turf. If you're working on retroviruses, then you're being judged on retroviruses. And if you break loose valuable men and women to hunt for this mystery disease which you know in all likelihood will turn out to be something other than a retrovirus, then you've got a potential problem. You may have helped the mycoplasma folks in the bacterial diseases division, if that's what it turns out to be, but you're not going to get any credit for that come budget time; they are. All you're going to show is that you haven't made as much progress on retroviruses as you needed to.

There's a hard but basic truth we all have to accept or actively fight to change, even in the public health arena and increasingly among individual care practitioners: all of us, every life form on this planet, from the simplest viral particle to the common cockroach, *Periplaneta americana,* to the plague-carrying *Rattus rattus,* to the African green monkey, to the harried government bureaucrat worried about his budget, to the most lauded Nobel laureate—we're all fighting for our own share of the resources. And the one thing nature tells us over and over again in the clearest of terms is that there is no good or bad in this struggle, no right or wrong, only winners and losers.

Since a bacterial infection was high on everyone's list, Rob Breiman, who was chief of the Respiratory Diseases Branch of the Bacterial Division, became head of the team and they de facto led the field investigation. Part of the decision had to do with personality as well. Mitch Cohen, who headed up the bacterial division, was an M.D. and former EIS officer with a strong epidemiology field orientation and little laboratory experience. So he was much more interested in sliding down the fire pole and racing off with the truck than my boss in the viral division, Brian Mahy, a Brit with a Ph.D. and a strong track record in molecular biology. He feels that the field folks are out there to provide samples and statistical confirmation for the lab scientists, who will use high-tech science to really figure out what's out there killing people. That, in a nutshell, is the culture clash between the field and the lab.

Within hours, Jay Butler was on a plane to Albuquerque with EIS officers. Joe McDade, the microbiologist who isolated the *Legionella* bacterium and who was now deputy director of NCID for laboratory science, soon followed.

We didn't need to lug along a field lab, as we usually did with outbreaks in remote foreign locations. The state people were first-rate and they were perfectly prepared to collect and make ready all the samples to be sent back to Atlanta for specialized analyses.

On the plane out, we talked about the cases and what we might use as a case definition—the collection of findings that would determine which cases brought to them would come under the heading of this disease. Then they listed their top candidates for what the mystery disease might be. Three-fourths of the original group of patients were dead. Now, it's a fact of epidemiology that many "new" diseases seem to start off with very high fatality rates. This isn't necessarily because they're more dangerous when first reported than later on, but because only the most serious cases will be recognized at first as belonging to the new syndrome. The milder or less clearly defined cases will likely escape detection or correlation. Of course, sometimes new diseases turn out to be worse than the initial indication. For example, Rift Valley fever virus was thought for decades to cause only a modest febrile illness; as we'll discover, that virus can also be a killer. What would we find in Four Corners?

One of the CDC team's jobs would be to try to work with the frontline folks to identify and investigate the entire clinical range of the disease and get those answers. It's pretty difficult to find a prevention or cure for something unless you first figure out what it is you're dealing with.

The world the team touched down in was a vast and eerily beautiful land of desert, scrub brush, deep canyons, and angular sandstone mesas sculpted by aeons of wind and erosion; our image of the true West shaped by generations of artists and filmmakers.

As they'd been trained to do, the team hit the deck running, immediately plunging into meetings with state health authorities and the clinicians who had cared for patients, interviewing survivors and their families, reviewing patient charts and hospital records. Press management is also extremely important, particularly when there are five separate health entities involved, and if they're not all telling a consistent story, the effects on public information and confidence could be disastrous. The reports coming in were still devastating and poignantly tragic. The latest person to die had been a thirteen-year-old girl who collapsed in front of her horrified mother at a school graduation party at Red Rock State Park.

The local authorities were impressively cooperative, particularly the New Mexico Department of Health and the Office of the Medical Investigator, who were real pros. But in addition to the epidemic, we realized we had to deal with something else that was just as important from a public health standpoint as the job our field epidemiologists were doing, and that was the handling of information. The media was all over the outbreak and wouldn't let it go. The terrified public was being inundated in a flood of conflicting information. Some stories seemed to say that everything was under control. Others were predicting imminent apocalypse. Neither was accurate or helpful to our efforts. Everybody working on the epidemic or even peripherally involved had an opinion and was ready to pass it along. Four state departments of health, the University of New Mexico, private practice physicians, and the Indian Health Service all had ideas, all wanting to be heard, since it was their turf. We needed a well-conducted Strauss waltz to inform and calm the public, but what we had was a simultaneous competition among heavy-metal bands. CDC dispatched a talented public information officer, Bob Howard, out to Four Corners to help manage the information flow. His task was to try to get everyone to sing off the same sheet of music. We wanted everyone getting out one message, and that message was: we don't know what this is, but we've got a crack team on it and they're working as quickly as is humanly possible. As soon as they know anything—good or bad—we'll make it public.

Discrimination against Native Americans by bigoted and frightened citizens continued to increase and there was more bandying about of terms such as "Navajo disease" and "Navajo flu." Health and Human Services Secretary Donna Shalala was so concerned she clearly made it known that government employees and public health officials would be responsive to Indian sensitivities.

High among these was the Navajo taboo against speaking of the dead or even mentioning their names for several days after death, until their souls had safely made the journey to the next world. The religious prohibition, however, didn't stop the press from trying to interview the bereaved families or publishing details wangled from confidential medical records. Reporters wanted to know everything, from what the victims had eaten before taking sick to intimate and intrusive questions about their sexual habits.

Some reporters were run off the Navajo reservation at gunpoint by tribal members. Since our people also had to conduct interviews and in some cases ask the same kinds of sensitive questions, the press had the potential to wreck the entire investigation. Many of the Navajo were also extremely uncomfortable with the idea of autopsies, since the traditional belief held that disturbing a body would cause the soul of the dead person to haunt the living.

And some of the older medicine men pronounced the disease a sign from the spirits, punishing the tribe for straying from the old values. Medicine man Ernest Becenti told one reporter, "There is a hole in the sky, and bad things are pouring through it. It's like the world has a hole in its roof that we caused."

The Navajo had been struck an incredible blow among their small community and had responded individually and through the leadership of their president, Peterson Zah. The legislative branch, the Tribal Council, was brought in to form a closer liaison between the many non-Navajo agencies working on the problem and the community in the long term. This was going to be a long struggle fought both on the reservation and off Indian lands in the rural areas of three states and we would need the resources of the Navajo Nation just as they would rely on us for our special expertise. The consensus was to prove critical for success.

Perhaps the greatest single talent an epidemiologist can have is the ability to listen. The tribal medicine men and traditional healers were spiritually attuned to their physical environment, were keen observers of nature, and came from a strong tradition of storytelling and oral history. And they told the medical investigators a consistent and fascinating story.

Their forebears had seen this pestilence before, they insisted, not just once but several times. And it had always been connected with an exceptionally long and rich piñon nut harvest, which had happened only three times in the last hundred years: 1918, 1933–34, and now, in 1993. Normally, the large edible seeds of the tree—also called the pine nut—are available for only a few months of the year. But in these three instances, the harvest had gone on practically the entire year. Along with this, the medicine men noted, had come an overwhelmingly large number of deer mice.

The deer mouse, or *Peromyscus maniculatus,* is common throughout much of North America. Its relative numbers are largely determined by how much food is available at any given time and place. This food is not limited to pine nuts. But many times in years when pine nuts are plentiful, seeds of grasses and other plants are also abundant. By virtue of the animal's short gestation period and the average size of its litter, the mouse can quadruple its population in about nine or ten weeks!

Meanwhile, back at CDC, I was directing the other part of the investigation, processing the samples that came back from the field. The people at CDC were in continuous contact with the people in the field, trying to give them ideas about questions to ask and answer and about samples that were needed and how they should be collected. The samples started coming in on Monday, May 31, Memorial Day.

If those of us in infectious diseases had our way, whenever a patient died mysteriously, a complete autopsy would be performed. Several small chunks would be taken from each organ and put in formalin jars with tight screw-on lids. Several more chunks would be placed in little vials for freezing, labeled with identifying numbers which would correspond to numbers on a packing list, then the whole package would be sent to CDC.

What we usually got was more like a big Styrofoam box packed with those blue ice packets and poorly labeled samples in plastic jars—the kind you give urine in at your doctor's office. So we had to call and say, "Look, this was sent from such-and-such hospital. Who sent it and what was the patient's name and when was it taken?" And the clock is running and people from all over NCID are beating on our door, yelling, "Where are my samples? I want to start testing!"

This isn't true of everyone, of course, and the New Mexico officials in particular were terrific. But I'd have to say that they represent one end of the spectrum, which stretches a long way in the other direction.

Pierre Rollin, head of our pathogenesis section, was anointed as the point man in processing the samples from the Four Corners area.

Pierre's a triple threat: lab, "epi," and a physician with extensive field experience. He's as well qualified as anyone to see the big picture.

We decided to process samples and run most of our tests at Biosafety Level 3. We could have gone directly to Level 4, since, in fact, we didn't know what we were dealing with, and that would have given us the highest possible level of assurance. But there were problems with this. First of all, it greatly limited the number of people who could work with the samples. Second, the Level 4 lab was full of Lassa and Ebola and all sorts of other truly scary stuff for which we have no cure. The chance of cross-contamination was small, but not nonexistent. Also, once we started at Level 4, we would have to stay at Level 4: and after being processed, these samples needed to go all over CDC.

So Pierre and the other scientists donned plastic bunny suits—white coveralls with a hood—respirators and eye protection. They worked under laminar flow hoods designed to keep any particles suspended in the air around the sample from getting out into the room. The room, in turn, was negatively pressured to keep everything inside it from getting out.

Pierre took the tissue samples and made homogenates out of them—grinding each one with a mortar and pestle, just as it's been done for decades. There's some danger in this, because you're potentially generating aerosols, and if that happens to be the means of transmission of this unknown agent, then you've got trouble.

The next step presents even greater potential danger. Once Pierre made the homogenates and put them in test tubes, he had to centrifuge them to get rid of the big chunks. The centrifuge spinning around generates a tremendous amount of energy, so anything on the surface of the test tube is aerosolized. And if the tube happens to break, huge amounts can be released into the air.

Somewhere within those samples would be lurking the bacterial, viral, or other kind of microorganism—a biological agent—that was making people sick. Our precautions should keep the genie in the bottle.

Once the samples were sent out to the various lab groups at CDC, everyone worked with them at the highest level of biosafety they were normally accustomed to using. For example, most bacteriologic work is done at Level 2 or 3, so the bacteriology labs analyzed their samples at Level 3. If they were culturing for a bacterium and, let's say, the agent turned out to be a virus, then the bacteriological medium they used in their petri dish wouldn't culture it and it would die out. But if it was a bacterium, then it could grow and the appropriate care would have to be taken.

In viral Special Pathogens, we often try to grow unknown microorganisms. It's a potentially dangerous strategy, because if you're successful, you're going to end up with a lot more of the stuff than you started with. We also inject them into animals, which is absolutely the most dangerous thing you can do with these viruses in the lab. As a result, we decided to do all animal work and most of our own examinations at Level 4.

Animals can be aggressive and unpredictable when they feel threatened, and though we strive to treat them as humanely as possible, they definitely feel threatened when they're kept in rows of cages and we come at them with needles. Once they're inoculated, if the organism grows in their bodies, then they begin excreting it in uncontrolled fashion, just as they would if infected in their natural surrounding. This means that for many of these viruses, you've re-created your epidemic right in the lab. We began by inoculating newborn mice and several other species of rodents. Newborn mice are particularly susceptible to viruses and their immune systems are immature.

Meanwhile, the bacterial lab people were trying to grow out their own samples. Some bacteria will grow on garden-variety culture media. Some require a specialized medium. Some have never been cultivated. So the bacteriologists were also undertaking some specialized genetic techniques that involve extracting DNA and doing a polymerase chain reaction test, or PCR. They also looked for more primitive types of bacteria such as mycoplasma and chlamydia, which don't have well-developed cell walls, making them fragile and pretty tricky to grow, so tricky in fact that we sent a selection of samples down the road to Gail Cassel's laboratory at the University of Alabama for special treatment. But they can both cause serious disease in humans.

In Special Pathogens, we began testing about twenty-five of the numerous virus strains we keep on hand against blood from victims of the mystery disease to look for immunological responses that would cross-react with any of the viruses we were working with. We took some from every possible category—arenaviruses, filoviruses, hantaviruses, nairoviruses, the whole shooting match—even though none of the agents we brought out were known to produce the specific symptoms that had killed the victims at Four Corners.

Of all the members of the biological world, viruses are among the strangest. Unlike the much larger bacterium, which can be said to have an independent life of its own, always in homeostasis, a virus can do nothing on its own outside a living host cell. We still argue about the origin of viruses and many believe a virus is a piece of DNA or RNA that broke off from a cell and learned how to spread to new cells. A virus may have fewer than a dozen genes, but that can be enough to make it fatal to an animal or human being with tens of thousands of genes. Different types of viruses have characteristic shapes, which is one of the ways we classify them. An arenavirus is a rough-surfaced ball. A filovirus is a long, slender thread that curves around itself. Rabies, which is virtually always fatal to humans without timely vaccine intervention, takes the shape of a bullet.

A virus has only one mission, one goal, and that is to replicate itself. It does this by attaching to the surface of the host cell, which it then penetrates, taking over the cell's reproductive "machinery" to turn out more and more copies of itself in its ongoing spread within the host body. Eventually it must make its way to the next host, and there are many strategies that it can take: some viruses like measles must be continually hopping to another susceptible human; others like herpes viruses cause latent infections and pop out intermittently to guarantee their "immortality." Most of the viruses described in this book have a reservoir outside humans, so they play out their drama of survival apart from our species. Like higher animals, viruses are subject to the laws of natural selection. Any virus that manages to survive for some time has probably adapted pretty well for its own needs.

Viruses fall into a strange netherworld. My colleague the well-known virologist Fred Murphy describes viruses as operating at the edge of life. In fact, the classic tree-of-life chart doesn't have a place for viruses. They're what's known as polyphyletic—they have no common ancestors. Some can be categorized and put into families on the basis of their shape or the way they reproduce, but they didn't all evolve from a single entity. Scientists have speculated they're simply pieces of someone else's DNA or RNA that somehow got loose and "fell off the tree." All viruses are pieces of either DNA or RNA held in a protective wrapper. The DNA or RNA is the genetic code that allows it to make more of itself once it gets to a host, and the wrapper is designed to protect it and get it into the cell it is going to infect.

Viruses get inside a host cell and use that cell's machinery to put their DNA or RNA to use making more virus. Darwin would have loved them: they exist only to make more of themselves and they are highly interactive in the natural selection process. With nothing to do but replicate, they do a lot of it quickly, which means there's plenty of opportunity for some genetic code (the DNA or RNA of the virus) to get botched in the process, creating what we call a mutation. And since they're always reproducing on someone else's home turf, any mutation that might make the virus more capable there is "selected for," allowing viruses to evolve and adapt much more quickly than their hosts can. Considering that some viruses reproduce in a matter of hours, even though the largest virus is 10,000 times smaller than anything we can see with our eyes, they are pretty formidable opponents.

In fact, the question might be: Why hasn't a species as big and slow to evolve as humans not been completely overrun with viruses and long since died out already?

First of all, not all viruses cause disease—much less death—and of the ones that do, not all cause disease in everybody who is exposed. Second, in the limited observations that modern science has made, we have not documented that viruses have wiped out any species, although they have caused epidemics that have decimated various animals and plants.

Fortunately for us, humans are not the main target for many viruses, which can be very picky, waiting for that specific host of choice that will allow them to thrive. Viruses that affect other species in the animal kingdom may have no effect on humans, or the effects may be different.

For example, monkey B virus—*Herpesvirus simiae,* the natural herpes virus of rhesus monkeys—causes a severe and usually fatal encephalitis in humans. Unlike viruses like measles, which survives in nature by jumping from host to host, herpes viruses operate under a latency strategy for survival which allows them to become inactive or latent and then reactivate in the same host over a long period of time—as anyone prone to cold sores can attest.

During the mystery disease epidemic, people all over the Center for Infectious Diseases, or CID were working virtually around the clock on the samples, testing them against everything we could think of. If it was a virus, and it was from a family anyone had ever seen before, the division of viral diseases wasn't going to take a chance on missing it. The bacterial people and the toxicologists were doing the same thing. Altogether, roughly seventy-five people at the CID were working in the labs on the outbreak, and another twenty-five or thirty out in the field, complementing the Atlanta efforts.

The guy in charge of our diagnostics in Special Pathogens was Tom

Ksiazek, who headed up the branch's Disease Assessment section and preceded me to Atlanta from USAMRIID. But ironically, on the first day of testing he was back at Fort Detrick mapping out a joint army-CDC project to test a hantavirus vaccine in Russia.

The U.S. Army first became aware of hantaviruses during the Korean War, when challenged with a disease simply referred to as epidemic hemorrhagic fever, or EHF, and later as Korean hemorrhagic fever. More than 2,500 American soldiers fell ill with this mysterious disease, which causes acute high fever, shock, body aches and weakness, and kidney failure. A hundred and twenty-one died.

It's a fact of epidemiology that most new diseases turn out to be old diseases. The Chinese had described a condition resembling EHF around 960 A.D. and from time to time in the thousand years thereafter. Russian scientists first noted it there in 1913, with sporadic outbreaks popping up in the far eastern Soviet Union throughout the 1920s. Japanese and Soviet researchers described what sounded like the same thing occurring simultaneously in Manchuria and Siberia in the 1930s, concluding that wild rodents were probably the reservoirs for this infectious agent. The Swedes had noticed a similar but milder syndrome, which they called nephropathia epidemica.

But it wasn't until 1978 that Dr. Ho Wang Lee and Pyung Wu Lee of the Korea University Medical School, working with Karl Johnson at NIH's Middle America Research Unit (MARU) and later CDC, finally isolated the virus in the lung tissue of infected *Apodemus agrarius* mice. They dubbed it Hantaan virus, after the Hantaan River, which flows through the DMZ near the site where it was first positively identified. At USAMRIID, George French coaxed the virus to grow in cell culture, an advance that allowed Joel Dalrymple and Connie Schmaljohn to study its structure and genetic material, resulting in the definition of a new genus called *Hantavirus*.

It's traditional among virologists to name a virus after the location of discovery. Not surprisingly, a lot of cities and towns aren't especially grateful to have their names forever associated with a deadly blight. Their chambers of commerce don't always consider it the best advertising. In some cases, then, a compromise has been reached to use some less sensitive geographical designation; hence the use of rivers. Karl Johnson avoided a lot of flak by naming the deadly filovirus Ebola, after a river in Central Africa near both the Zaire and Sudan outbreaks.

Late on June 3, after a lot of people had finally gone home and most of the labs were dark, I went in to see how the diagnostic lab was coming along. Cuca Perez-Oroñoz, one of our lab technicians, had just been reading her fluorescent antibody screen results of patients' blood sera, looking for an antibody cross-reacting against any of our battery of various viral antigens. The arenaviruses were negative . . . negative . . . negative. Just as we expected. Then she ran off hantavirus antigens including Hantaan and Seoul viruses.

Positive . . . positive . . . positive.

Then she tried other viral antigens. All negative.

I went across the lab to Mary Lane Martin, who was running the ELISAs. ELISA, short for enzyme-linked immunosorbent assay, is a blood test with many important applications, including screening blood donations for various infections.

I asked Mary Lane if she'd run any hantavirus ELISAs.

"No," she replied, "I ran the arenaviruses first because you told me to work through all of them."

"Okay," I said with some urgency, "but be sure to run the hantas tomorrow."

The next day, while Pierre Rollin, an excellent diagnostician, and I were down there, she did. And she discovered several serological positives against hantavirus. The ELISA isn't like your standard biochemical or hematological blood test that gives absolute values in milligrams or micromoles. They still have to be interpreted by someone who's had a lot of experience with the virus, and these tests were Tom Ksiazek's babies. When Tom and I had been at USAMRIID, he'd gone to Yugoslavia, where there were three different hantaviruses circulating, with a test for one of them. He had seen what the cross-reactions looked like and how difficult they were to interpret and he had since set up tests for the other Yugoslavian hantaviruses. So he knew exactly how these tests performed in the real world and not just in laboratory conditions.

Pierre made a copy of Mary Lane's results and faxed it to Tom at Fort Detrick. When Tom read it, he hurried back to Atlanta.

Tom, Pierre, and I were in a quandary. We thought we had the data that made the serological connection and felt confident of the data because these were the same reagents that we had developed and tested at USAMRIID. But we still had to face the fact that we were working in the United States—and there had never been any evidence that hantaviruses here caused any disease at all. How does a bug from Asia, known to be associated with rats and mice, make its way to a landlocked region thousands of miles away? In Europe and the Far East, hanta had *always* been a renal syndrome—it devastated the

kidneys—*never* a pulmonary syndrome. So even if we had detected
this particular virus, it could be an incidental finding, unrelated or
only vaguely related to what was actually killing the people in Four
Corners. If *The New York Times* announces on page 1 that we know
what's going on and we turn out to be wrong, the confusion would
be terrible and the effort to control the disease would be set back,
perhaps for months. Further, the damage to CDC's credibility would
be enormous. This is not simply institutional hubris. Much of CDC's
effectiveness derives from the delivery of reliable, unbiased information to the public.

As soon as we cautiously presented our preliminary findings to our
CDC colleagues, we got the reaction we expected: *So you found
some hantavirus antibodies in some of the samples; it's an irrelevant
finding. It just happens to be there. The disease is really nutritional,
or it's social, or it's multifactorial—a chlamydia plus a hantavirus
plus maybe something else you haven't even found yet. The point is,
C.J., hantavirus causes renal problems, not respiratory problems,
and we don't see it as a clinical disease in the United States.*

At this point, we hadn't brought our molecular biologists into the
hunt, even though their state-of-the-art genetic sequencing techniques were theoretically the most precise and accurate. Polymerase
chain reaction, or PCR, is, in a sense, a way to amplify extremely
small quantities of the genetic code so we can analyze it in detail.
First, we have to make a DNA copy of the RNA virus genome we
suspect may be lurking in the tissues of a dead "mystery disease"
patient, using an enzyme called reverse transcriptase—so called because the usual way the genome code is transcribed is from DNA to
RNA. Then we try to selectively increase the amount of viral DNA
copies. This latter reaction uses another enzyme that makes many,
many copies of the DNA we are looking for. The DNA that is to be
amplified is targeted by using genetic sequences that are similar and
bind to it. When we have a goodly amount of the product, we can
find out its genetic sequences and compare them to those of known
viruses.

The problem is that while such approaches work well in controlled
situations in the lab, that's not necessarily what you get when you
bring in a piece of lung from somebody who's died and you're trying
to identify a possible virus in the tissue, which you don't know much
about.

We had to get to the bottom of this quickly. Already, people were
dead and the disease was not going away. We had an immunological

response to the hantavirus. We also knew that hantaviruses are hard to isolate or grow in cell culture, so we couldn't count on isolating it anytime soon. But now that we knew what we were looking for, it was time to bring in the molecular biologists and their PCR tests. With a little luck and a lot of skill, we might be able to get some specific genetic sequences that would identify the RNA, the genome of the virus.

It was something that had never been done before for a hantavirus. No one had ever found a hantaviral genome in tissues from patients, ever. But we also didn't think it had ever been tried seriously by anyone really good.

A unit like Special Pathogens at CDC is a lot like a baseball team. To be effective, you've got to have people with special strengths. And in the area of PCR and genetics, Stuart Nichol is at the top. There are few, if any, molecular biologists who are better at this than he is. Stuart also had some money left over from one of his grants, and was given permission to bring his remaining money with him to support his graduate students when he moved to CDC nine months before the epidemic. This isn't an ideal way to support research into a deadly virus, but it's a reality of modern life we have to live with in this era of personnel restrictions and budget cuts, and you learn to beg, borrow, or steal wherever you can to get the job done.

Stuart accessed the gene bank, a government-funded computerized database where we store all reasonable genetic information from everything that's been examined. Most scientific journals make depositing these sequences a condition of research publication.

With polymerase chain reaction, you need something called primers. Essentially, you have two gene primers that match up with similar sequences in the target and then you bridge those primers and copy what's in between, which then allows you to determine the sequence of what's in between and make some inferences about the genetics of the virus. Stuart designed two sets of hantavirus primers: one was to be sensitive to Hantaan and Seoul viruses from murid rodents and the other worked best with viruses like Puumala and Prospect Hill viruses from arvicolid rodents or voles.

The direct PCR reaction indicated we had something. We were all kind of surprised, though, that it seemed to come from murid, or Old World, rodent virus primers. But we weren't about to look a gift horse in the mouth.

Stuart's a careful guy. We didn't have an automated sequencer, but the influenza section did. So Sergei Morzunov, one of Stuart's non-

CDC postdocs, slipped over there at night when it wasn't in use to do the sequencing. When he fetched the data and compared it to the gene bank the next morning, it turned out to be a false positive, a normal cell gene, deflating our hopes considerably.

Stuart then ran a second amplification called nested PCR. And this time he hit pay dirt. We had concrete evidence of a hantavirus in the tissue samples from Four Corners. But it was a *new* hantavirus, something never before seen by medical science. No wonder the symptoms didn't match up. We later named the disease it caused hantavirus pulmonary syndrome, or HPS, to distinguish it from the hanta disease we all knew, hemorrhagic fever with renal syndrome.

Now we were on a roll. One of our researchers, Suyu Ruo, had made some monoclonal antibodies against hantaviruses and she had one that seemed to react to all of them. A very talented pathologist we worked with, Dr. Sherif Zaki, took it to Building 7, down to the sub-sub-basement, where he used the hanta antibody to search for hantaviral antigens in the tissues. And he found them overwhelmingly in the lining of the tiny blood vessels in the lung.

We now had an immune response from field samples that reacted with hantavirus, we had hantavirus genetic material in the tissue, and we had localization of hantaviral antigens in the right tissue to explain the clinical syndrome. That didn't mean we had a cure. But we had clearly identified the enemy, and in medicine, that is always the first step.

Knowing the disease was caused by a hantavirus told us how it was spread as well. Hantaviruses are carried almost exclusively by rodents and transmitted to humans through tiny droplets of urine or bits of feces freshly shed or in the dust and dirt that get stirred up and inhaled. This squared perfectly with the reports of the Navajo medicine men who directly correlated previous disease outbreaks with abundant piñon nut harvests and a corresponding explosion of the deer mouse population. Jim Cheek had been extraordinarily lucky to avoid exposure during his initial investigation at Merrill Bahe's trailer.

Our first instinct is to yell "Bingo!" and announce it to the world. But we don't, because good science is based on peer challenge and self-criticism. Instead we kept on doing the tests over and over again with each new batch of samples and tubes of sera that came in, to be certain we were right. We injected blood from victims into animals in the Level 4 lab, got it to multiply and show an even stronger antibody reaction to hanta reagents. Nonetheless, it was the shortest amount of time ever needed to identify an unknown virus during a disease outbreak. Legionnaires' had taken six months of concentrated effort.

There was still some skepticism about our results. At a CDC staff meeting I told those assembled that the way we would finally identify the exact strain of hantavirus was to isolate it in the disease reservoir itself—the deer mice or other rodents we suspected of transmitting it.

Jim Hughes, the head of CDC's Center for Infectious Diseases, was in a tight spot. The outbreak was all over the national press and he was the guy who had to manage it. His background was in hospital infections and bacteriology. He also came from an Epidemic Intelligence Service background. They're the first ones out in the field, so their article of faith is that epidemiology solves the mystery and then the lab merely confirms what you already know. It's my belief that epidemiology gets you the right clues and samples and the lab solves it. In fact, during the Four Corners outbreak I lost it one day over one of our guys in the field who was spending all his time reviewing hospital charts rather than shipping back the samples we'd sent him out there to get.

So Jim has a lot of people like Mitch Cohen, head of the bacterial diseases division, and other former EIS officers without laboratory training who weathered storms together with Jim, telling him hanta must just be a co-factor. On the other hand, he's got me, this former army colonel from USAMRIID they just hired the year before who's telling him it's a hantavirus and it's time to get moving, because the rodents are pissing and the humans are breathing the droplets and dying.

If Jim announced our findings to the press and we were wrong, they would have crucified him. But if he held back, he'd get nailed for withholding information. A high-risk enterprise all the way around.

Jim went through all the data himself, carefully reviewed it, and was convinced. Wisely, he also decided to seek outside counsel and buy himself a little insurance and insulation in the process. He assembled three guys with big reputations: retired General Phil Russell, M.D., formerly the army's top man in infectious disease research; Fred Murphy, D.V.M., Ph.D., who had been head of CID before Hughes; and Scott Halstead, M.D., at the Rockefeller Foundation, one of the pioneers in dengue fever. These were three smart guys who'd been around the block a few times. They're sort of grandfather figures in the world of infectious disease research.

We had stuff faxed to them to read while we presented our findings and then asked them in the course of the conference call, "Well, what do you think?"

To a man, they all said, "Looks like a hantavirus."

In fairness, all three of these guys had had large roles in my own training along the way, so I had a pretty good idea of the way their minds worked.

So at that point, Jim was ready to release the information, which we published on June 10 in CDC's *Morbidity and Mortality Weekly Report,* or *MMWR,* the publication elaborating all the outbreaks and diseases and other bad things that have happened to people in the past week.

The editor of *MMWR* was Rick Goodman, another veteran of the Epidemic Intelligence Service. Back in 1978, on his first case as an EIS officer, Rick was sent out to the Midwest to investigate what was reported to be an epidemic of respiratory disease. When he arrived, he found that it was really just several cases of unrelated problems that happened to occur around town at the same time. But the index case was one that made a tremendous impression on him: a young, vibrant father of two who suddenly had trouble breathing, developed severe pulmonary edema, and died in the hospital despite intensive medical support. It was a terrible tragedy.

We often joke that there's no one smarter than the guy or woman who's just finished residency, at least they think so. And it confounded Rick that he couldn't figure out what had happened to this man. The autopsy turned up no specific cause. He kept in contact with the man's young widow. He kept the old file from the investigation for fifteen years. The case continued to weigh on him.

Then, as he edited the reports on hantavirus pulmonary syndrome, he stopped and said "Wait a minute!" to himself. He pulled out the old file and went through it again.

"Looks like HPS to me," I said when I saw it later.

He tracked down the man's wife again, got her permission to have the hospital release tissue samples, and had them shipped to CDC, where Sherif Zaki stained them. There it was in the tissue. As had happened with Legionnaires', a medical mystery was solved more than a decade later.

Once you think you know what you're dealing with in a disease outbreak, the next step is a two-pronged strategy of treating the sick and eliminating the source.

There are many antimicrobial drugs that work well against bacteria, but these antibiotics have no effect on viruses. There are, however, a small number of drugs that are very effective against viruses, such as acyclovir against herpes viruses, or the expanding arsenal against HIV. The only such weapon in our arsenal against hemorrhagic fevers was ribavirin, which seems to be highly effective against arenaviruses such as Lassa fever or Bolivian hemorrhagic fever. We immediately threw it into the struggle against hantavirus pulmonary syndrome because of suggestive results with the Asian hantaviruses.

The other avenue of attack was the source. A team composed of CDC people led by Jamie Childs, Navajo and other health officials, and the biology department of the University of New Mexico, now properly protected with masks, goggles, gloves, and paper body suits, braved the 100-degree-plus heat to get to every site where there had been a report of illness or death and trap as many small mammals such as mice, rats, prairie dogs, rabbits, and skunks as they could.

Within a few days, we had the information that confirmed our suspicions. Rodents were everywhere, particularly in case households, and they tested positive for hanta. *Peromyscus maniculatus,* the common deer mouse, a cute little species that looks as if it could star in a Disney nature film, turned out to be the most prevalent reservoir and 30 percent were positive. By the middle of June we'd sequenced viral genetic tissue extracted from captured deer mice and found it to be identical to viral genetic material from human patient tissues. This also gave us a clue as to why the disease was happening now. The traps were full of mice, and comments by students of rodent biology in the western states could be summed up by one's pithy statement: "The rain and weather are good, the grass is high, and the rodents are everywhere, and the coyotes are sleek." The virus had been in deer mice all over the United States, waiting for the kind of weather that allowed food sources for the deer mice to explode.

We don't think the ribavirin did much good in treating the sick; we're not sure. The mortality rate did go down, but that could have been due in part to a wider recognition of less serious cases and a greater sensitivity to the disease on the part of primary-care docs, meaning that patients were sent to the hospital sooner for quicker and more aggressive supportive care. The mortality is still 40 percent and we have a long way to go.

Within a month after announcing our findings we'd examined more than 10,000 samples, and by the end of July, CDC published a set of recommendations for risk reduction. By late fall, scientists at both CDC and USAMRIID had independently grown HPS viruses in cell cultures. What we had never suspected in the busy days of June was that the workload would continue without any let-up or additional staff to help. We got some dollars, but the bodies didn't come, so we suffered the inefficiencies of short-term people and tried to keep morale up. The additional dollars bought computers and automation to help our efficiency, but what we had really hoped for was staff to share the workload; core staff had been in the lab all summer on twelve- and sixteen-hour days, seven days a week. When Tom Ksiazek's wife and daughter wanted to see him, they brought a picnic lunch to eat in his office on a quiet Sunday evening.

Of course, it wasn't possible to eliminate the entire population of southwestern deer mice, and even if it were, species genocide is not a business any of us ought to be in. The only realistic approach to decreasing incidence of the disease was to try to decrease rodent-human interactions, cleaning up buildings or areas with heavy rodent infestations, and taking precautions with activities known to increase the risk of contact, such as planting or plowing or other field work. With CDC's help, the state health departments put together pamphlets, posters, and videos in English, Spanish, and Navajo demonstrating how to avoid contamination, which were distributed to schools, libraries, churches, and companies employing workers likely to be in contact with potentially disease-carrying animals.

Although we confirmed forty-eight cases of hantavirus pulmonary syndrome in the Southwest, including twenty-seven deaths, we thought that represented only a small percentage of the actual number of cases. In the furor surrounding the epidemic in the Four Corners, additional cases from the same virus were found in more than twenty-five states in the United States, as well as in Canada, and they continue to be reported every year, although in lower numbers. Indeed, during the search two other lethal human hantaviruses were found in the southeastern United States. The clinical disease in the Southwest was carefully described by Jeff Duchin, Fred Koster, Rob Breiman, and others on the team, and it has turned out to be very distinctive. So we have not just a "new" virus but also a "new" syndrome. This has led clinicians to suspect the presence of hantavirus pulmonary syndrome in Brazil, Paraguay, and Argentina, and we have been able to confirm that these outbreaks are caused not by the North American viruses but by other, related viruses with their distinctive rodent reservoir and thus different epidemiologies.

Research into hantaviruses and more effective ways of controlling and treating them continues. Like almost everything else in public health, this battle remains unfinished and the victory incomplete. We are working with universities in New Mexico, Arizona, and Colorado to monitor the fluctuations of rodent population numbers, their infection with hantaviruses, and the relationship of these changes to ecological changes. We hope that someday instead of simply telling people to avoid rodents whenever and wherever possible, we hope we'll be able to say, "This year, in this place, because the weather is thus-and-such, there are certain activities you should be careful about."

Perhaps there isn't enough of this disease in any one place at any one time to warrant the effort to try to produce a vaccine, but I do think we should understand enough about the immunology of the virus so that if there's some drastic change in its properties or change in the ecology that increases its impact, we'd at least be in a position to begin working on one experimentally. In the Patagonian region of Argentina during 1996–97, hantavirus pulmonary syndrome surfaced again. There was an outbreak of twenty cases, mostly in the small town of El Bolson. This attack rate (one case per thousand people) of meningococcal meningitis would be a severe epidemic and in the Americas would trigger a panic and an emergency vaccination program. We couldn't find the galloping numbers of rodents that made sense of the epidemiology in the southwestern United States in 1993. For the first time, we saw an epidemiological suggestion that a hantavirus might be transmitted from one person to another. We have to keep on top of this disease. A vaccine may well be needed in South America.

As a result of their experience with such outbreaks as the one in Four Corners, the New Mexico Department of Health has set up a multidisciplinary infectious disease death-review team. The group, which includes the Office of the Medical Examiner and the University of New Mexico, meets regularly to examine any death in which an infectious agent is suspected.

Our initial desire was to call the new agent the Four Corners virus. It turned out, though, that the Navajo were unalterably opposed to this name because the Four Corners intersection was, in their view, intimately associated with the Navajo reservation.

In November we had a meeting in Albuquerque, at which we formally announced the isolation of the virus to the scientific community. I suggested several possible names, which a number of us had trimmed down to a short list, and which I sent to the health director of the Navajo Nation. When she didn't answer my letter, I should have known there was a problem.

I tried phoning her several times, but she never returned my call. When I finally tracked her down, I said, "Look, I have to register the virus and give it a name. Do you have any problem with Muerto Canyon virus?" Muerto Canyon was a dry arroyo not far from the place where the first people had died.

"Well, no, I suppose that would be okay," she replied with polite circumspection.

And so in January 1994 we registered the name. Muerto Canyon: Canyon of Death.

In a stunningly short time the shit hit the fan. Some Navajos were up in arms. Representatives of the National Park Service were upset because they already had a location called Canyon de los Muertos— Canyon of the Dead Ones—as part of their Canyon de Chelly National Monument in Arizona and felt an association with a deadly virus could possibly be bad for tourism. Native American groups didn't like it because the Park Service's Canyon de los Muertos had been the site of an Indian massacre and they didn't want the enormity of that act overshadowed by any other association.

Ultimately, after going back and forth on a wide range of possibilities, we decided on the one name that nobody was completely happy about but everyone could live with. The Navajo Tribal Council approved it unanimously.

The designation? Sin Nombre virus: No Name virus. For the killer without a name.

2

West Texas and Beyond

I SHOULDN'T have been surprised by the tussle over naming the Four Corners virus Sin Nombre. In one way or another, I've always had trouble with names: both remembering other people's and trying to forget my own. Although most of my friends and associates call me C.J. or Pete, I came into the world as Clarence James Peters, Jr., on September 23, 1940, in Midland, Texas, a town in the western section of the state whose wealth was created by the oil industry boom. They used to say that if you walked into the lobby of Midland's Scarborough Hotel any time of the day, you'd always see at least one millionaire there. In fact, I'd say there were two kinds of people in the area in those days: those millionaires, then all the people who worked for them, either directly or indirectly.

My family belonged to the group who worked for them—indirectly. My father, Clarence James, Sr., was of German stock from a German-settled town in Texas, New Baden. My mother, the former Helen Lorene Webb, was from a mixed Scottish-English heritage. Since my dad was already known as Pete, I was called Little Pete or, sometimes, Re-Pete, until I got older and was able to share just plain

Pete with my father. Whenever I heard myself referred to as Clarence James at home, I knew I was in trouble. And the only time I'm called that now is by the IRS or some salesman who's gotten my name off a computer list.

By the time I was in first grade, we'd moved to Odessa, which is only about thirty miles down Interstate 20 but much more blue-collar than Midland. Odessa was where the drillers and the "rough-necks" (the term for someone who worked on an oil rig under the driller) lived. My father, who was foreman of an automobile repair shop, was outside the oil business, but anyone who lived in the area supported the industry in one way or another.

Even with 50,000 residents, Odessa was a small town. The only time you'd see a hint of traffic was right around five o'clock in the dead center of downtown. It was a safe, easy place to grow up.

Although there weren't any other children in my household, we were lucky enough to live next door to an empty lot, which meant there were usually plenty of kids nearby for games of hide-and-seek, kick the can, and work-up softball. We seldom had enough players for two full teams, so we'd take turns, starting in the outfield after being put out at bat. I was neither a great athlete nor a terrible one— neither the first guy chosen nor the last—but I was enough of a realist that I never entertained fantasies of a big league career.

Maybe because I was an only child, I was always close with my parents; no major clashes or conflicts of values. I was also close with my maternal grandparents, who owned a dairy in San Angelo, about three hours' drive from us—at West Texas speeds. Along the way you'd see oil rigs dotting the horizon, pumping the black gold like giant birds pecking at the Texas dirt.

We used to visit my grandparents several times a year and I would spend summers there. During those summertime visits I learned to spend a lot of my time entertaining myself. My grandfather would be out at the dairy, my grandmother would be busy in the kitchen, and my aunt, who was a dressmaker and lived with my grandparents, would be cutting and sewing or seeing her customers in a large, attached room. So it was up to me to amuse myself.

I'll never forget a set of Compton's Picture Encyclopedias in my grandparents' house. I would often sit all afternoon meandering through those books. The encyclopedia was our generation's version of hypertext. When I looked up Egypt, for example, I'd see a picture of the pyramids and be transported there in my own mind. I'm not sure how much of my later yen for travel and exploring foreign cultures stemmed from those afternoons with that set of Compton's, but there's no question that they opened up the world for me, far beyond the expanses of the flat Texas plain.

I also think that spending time on a dairy farm, seeing the way animals and humans interact, helped shape my attitude about the roles of animals and human beings in each other's lives. My grandfather doted on his animals. He treated them with respect and appreciation for what they provided him and his family. He was a perfectionist and was very proud of the fact that nobody took better care of their cows than he did. Milk from his cows was used as the starter milk for the buttermilk produced in the area.

In Odessa, we lived only about six blocks from the public library and I would go there many days after school, if for no other reason than to escape the oppressively dry Texas heat, which often topped 100 degrees and would give you a burning feeling on your back like baking toast. The library, by contrast, was air-conditioned and cool.

Religion played an even more important role in my upbringing. My mother's family was pretty easygoing Methodist and my father's people more intensely Baptist, the denomination in which I was brought up. My parents were moral and ethical people, though not big churchgoers. Sometimes they would go with me, but more often they would drop me off at the Sunday school. The most zealous churchgoers in my family were my father's parents, particularly my fire-and-brimstone grandmother. I remember being out with her in town one day and seeing the grocery store on fire.

"Serves them right," she huffed. "They sell beer."

My own views on religion, however, remained largely unclear to me. Even before I knew I wanted to study science, my nature led me to look for empirical answers to life's mysteries. How could you prove what was right or wrong, or be sure that God really existed? These questions continued to bother me, and I kept wondering when and where I could get to the truth.

One Sunday while I was in junior high, my parents dropped me off at Sunday school as usual and it turned out to be one of those "Come to Jesus" affairs they had with some frequency. They would play music and everyone would put their head down and pray as the leader prompted, "If you've taken Jesus for your savior and you want to come down here and profess it, then come on down!"

As this was going on, the visiting minister asked perfunctorily if there was anyone in the audience who had any doubts. I very earnestly raised my hand. To this day I can't tell you what I was thinking. I just know I was immediately hot-boxed. Not only was I the only one that day who admitted having any doubts, I was probably the only one they'd seen in thirty or forty years to admit to such a moral morass! I was promptly put on the spot and asked what my doubts were and what they could do to dispel them immediately. Well, I couldn't explain what I didn't understand and therefore had no idea what they could tell me to make me believe. I've still never had one of those mystical epiphanies or conversions so many of my neighbors talked about.

But what I have found over the years is that science is as much a belief system as religion and each belief system is best left in its own domain, rather than trying to usurp the ground of the other. With both religion and science, we're talking about what we believe is true. And in both cases, if we're wrong, we may end up in a place the religious among us might call hell.

The other religion in West Texas, of course, was high school football, and on the night of the big game between Midland and Odessa high schools every year, the rest of the world stopped. The football players were the school and town heroes, and all the prettiest girls dated them.

Where did you go on a date? Usually to the movies. One of the less expensive options was the Lyric Theater on Saturday afternoon, where for twelve cents a ticket you got a double feature. But this was more of a place to go with your buddies than with a girl. If you had an actual date, you'd want to go uptown to the Scott, which was much classier and had a balcony. The only problem was, it cost fifty cents a ticket. Not a lot of options in a place like Odessa. Still, if you could take a girl to the Scott, you had a much better chance of getting her out to Sand Hill State Park afterward, which was secluded and romantic and, in the 1950s before easy contraception became available, both an exciting and a risky venture.

The biggest influences on my life were my parents, even more than for many people, I suppose, since I was an only child. My father had gone to college for a couple of years until World War II imposed a different set of priorities. He'd always been in love with cars, tinkering with them, and in the army he became an automotive adviser, teaching people how to fix trucks. Once the war was over, he began working as an auto mechanic rather than going back to school. He was quickly promoted to shop foreman.

Looking back, my dad was an unusual guy in many ways, although I didn't know it at the time. He always treated people fairly and equally in a time and place in which it wasn't always the custom to do so.

Odessa wasn't exactly a center of ethnic diversity when I was growing up. There was a fairly sizable Hispanic population that had come up from Mexico, many of them one or two generations ago, but that was about it. There weren't many blacks in West Texas; those who were there were not only segregated economically and socially, they were also segregated in the workplace.

Back then, at a typical automobile garage, the only blacks you'd see were maybe one or two who were straightening up or cleaning the floor. My dad, however, always selected a couple of black guys he thought had potential and put them to work fixing things and doing small jobs at the shop, working their way up to larger jobs. Once they were trained, most of these guys would leave and find work as full-fledged mechanics or open their own garage—usually in a predominantly black community that would accept them at a professional level. It was kind of a losing proposition for my father. He knew he would lose a good man once the guy had acquired all the skills, and in doing so, he risked arousing the resentment of the whites, who sometimes felt my father had passed them over in favor of blacks.

But my father was never too concerned with what the community at large thought of him as long as he was comfortable with himself. I learned a lot from him about self-respect, self-confidence, and doing what you think you ought to do regardless of what other people think.

The worst and most searing experience of my childhood—and undoubtedly an event that, at least subconsciously, helped shape my career in medicine—was my mother's illness. When I was in junior high school she was diagnosed with lung cancer. She and I were very close. We all spent several painful years fighting the disease before she died during my sophomore year in high school.

My mother was incredibly brave. She always kept up a positive front around me and did her best to keep our family life normal. My father, as a result of my mother's illness, became less carefree and much more serious, trying to support my mother and me emotionally while he took over more and more of the day-to-day responsibilities around the house, at the same time that he was struggling to handle mounting medical bills on his relatively modest salary. As my mother's cancer progressed, I began going through a series of hard realizations. I began to understand that the woman who'd been with

me all my life might die—no matter how much my dad and I loved
her, no matter how much it hurt us and hurt her. It wasn't fair—she
was too young. It wasn't equal—other mothers weren't dying. It just
was the way it was.

As her condition grew worse, she developed an ulcer, which perfo-
rated at home. At that point, they did a gastrectomy. Then, because
of the steadily increasing pain, they did an endoscopic exploration,
looking in her thoracic cavity, finding metastases on the pleural surface. It was this that was responsible for the pain. Unfortunately, it
couldn't be adequately controlled with drugs, even strong opiates. So
her doctors decided to attempt a neurosurgical procedure to cut the
sensory tracts. It was a very dangerous operation, but at that point
we'd been through so much already, it was just one more terrifying
prospect.

It was the operation that finally killed her. Afterward, I rational-
ized that at least her suffering had come to an end. But in my more
introspective moments, I questioned what kind of just and righteous
God could allow such a wonderful person to be tortured so horribly
for so long. I became more fatalistic and, to some extent, have remained that way ever since. But mainly I remember an incredible
sense of loss and loneliness.

Did this protracted experience make me want to go into medicine?
Not consciously. I had an awful lot of contact with doctors throughout my mother's illness and it was all so unpleasant that I don't
believe I ever consciously thought, "That's what I want to do with
my life."

One thing I do know, however—once I became a doctor, my
mother's experience made me a better, more caring and compassionate one. Even as an intern and resident, I think I had a deeper understanding of my patients' problems and fears because I had been there
to see what she went through. For many physicians, especially early
in their careers, the patient-care experience ends once the patient
walks, is carried or wheeled, out the door. You've diagnosed the
source of their illness, advised on how to mitigate and/or control
their symptoms, and the rest is up to them. But before I ever took my
first patient history, I understood that from the patient's point of
view, the experience *begins* once he or she walks out your door and
has to go back to his or her job, or home to take care of the kids,
regardless of how that person feels or how weak he or she is from the
treatment.

The year of my mother's death I encountered another teacher who turned out to have a profound—even critical—influence on my future—my biology teacher, Mrs. Willie Filleman. From Mississippi originally, in her mid- to late-forties, short, with a sort of wide, roundish face that reminded me a little bit of a frog, Mrs. Filleman was a voluble woman who was always pushing her glasses up on her forehead. She was married to a very nice, slight, skinny little guy and it was always said he had something wrong with his heart. She was always so hyperactive that you couldn't tell whether he was quiet and sick or just couldn't get a word in edgewise.

She believed in differentiating, in a very clear way, the top of the class from the bottom. And every year she failed people who didn't make it, even football players, though in retrospect I suspect someone in the principal's office probably looked out for most of them and made sure they got in the school's other, far easier class. The lab benches we used as desks in her classroom were made of wood and painted black. During tests and quizzes I would sweat so intensely that the black residue would come off on my hands and stain the paper. She really put the pressure on.

But as hard as the class was, I found I really loved the subject matter. One day I went up to Mrs. Filleman and asked, "How do you type blood?" I can't remember why I was so interested, or what gave me the courage to approach her.

She said to me, "You want to know how to type blood?" She seemed pleased. "I've talked to a fellow at the hospital who works in the laboratory and he's willing to take on one or two students and teach them a few things, after school or on the weekends. His name is Zorus P. Colglazier. Go down and talk to him." She gave me his phone number at Ector County Hospital—one of the places where my mother had received much of her treatment—and I made arrangements to see him.

Zorus P. Colglazier was the head medical technologist at the county hospital. He was born and raised in Oklahoma and after high school he'd gone into the navy. The navy taught him medical laboratory technology and after he got out, he furthered his skill at the Gradwohl School of Laboratory Technology in St. Louis. When I met him, he was in his mid-thirties, short and muscular, with long blond hair and still wearing navy bell-bottoms.

I started out with Zorus pretty much where most people start in a clinical lab: doing urinalysis. Today, we use these little pretreated dipsticks to test urine specimens, but back then urinalysis involved a whole series of steps and a lot of chemistry you just didn't learn in high school. Although it wasn't very complicated, for a high school kid this was exciting stuff, and very different from working as a roughneck, the other summer job available to someone my age in Odessa. It was work that seemed important. How well I performed my tasks would actually matter to someone who was sick.

Before too long they let me start running more advanced tests like the EKG and drawing blood and performing the blood tests. Today the tests are pretty much automated, but back then we had to measure precise amounts of serum into a pipette, add reagents, heat the mixture, and then read and record the results.

The more I learned, the more thoroughly I enjoyed it. I decided I wanted to go to college and major in chemistry in the hope of getting into some kind of medically related research. If you didn't go to college, the likelihood was that you'd end up working on an oil rig, where, at any time, you could have a piece of pipe swinging at you across the rig floor, or undergo any number of other mishaps. If you worked hard, you could progress to driller, the guy who told the roughnecks what to do. That was basically the career progression in Odessa in those days and the prospect of spending my life on an oil rig was a powerful motivation to try for college.

Another experience that pushed me toward college was a government-sponsored program I participated in during the summer after my junior year. Developed as a reaction to the Soviet Sputnik launch and the national paranoia over our lagging behind in science education, the program sponsored "promising" high school students to take math classes over the summer in a university setting. Two or three of us from Ector County traveled to Houston to attend mathematics classes at Rice University. This gave me a taste of college life and independence, and it didn't cost my father a penny—very important for me since he was still working day and night to pay off my mother's medical bills.

The program was run by a fortyish math professor named Lincoln K. Durst. He was a fantastic lecturer and an inspiring teacher. He didn't have funny pictures or fancy computer programs, but he had the capacity to convey the intrigue and the beauty, the symmetry and asymmetry, of modern mathematics. We all loved him and his young blond wife, Claire, who was cute enough to stimulate our high school-sized libidos and still bridge the generation gap, becoming an older sister to us all.

When the time came to apply to college, I already had Rice at the top of my list. I had loved the summer experience there, and with the endowment originally provided by William Marsh Rice to create an institute "to educate the white men of Houston, Texas," there was no tuition. That original charter was gradually modified to include all races and both sexes, but at this point most of the students were still from in-state.

The euphoria of getting in, however, was tempered somewhat by the tone of our welcoming speech, delivered by Professor George Holmes Richter, author of the widely used textbook on organic chemistry. Richter was tall and imposing, with a shaved head, and always wore a blue serge suit. He welcomed all 400 of us freshmen by saying, "Look to the left of you. Look to the right of you. One of the three of you is not going to graduate. Don't let it be you."

Though I didn't know it at the time, this was probably my first lesson in the rule of nature all biomedical scientists have to grapple with. Organisms—both visible and microscopic—are constantly in competition with each other. Not all of us can occupy the same turf. The winners, as Dr. Richter cautioned us, have to earn their place.

There were a number of important courses at Rice that changed the way I thought about things. One, adolescent psychology, taught by a guy named Trenton Wann, who was also a clinical psychologist at the Veterans Administration hospital, showed me that none of the models used by psychology could grasp the essence of what a human being and human behavior are all about. Like most college kids, I was probably better at the academic discipline of adolescent psychology than the applied one, which meant I got into my share of social scrapes and broken hearts. But studying it in his classroom was all a marvelous introduction to the philosophy of science. What I got out of it in a useful, practical way was the realization that when you deal with other cultures and behaviors, you have to be able to understand what's going on from the vantage point of the other culture. If you're dealing with the head of an Indian tribe, or a village in the middle of the Central African Republic, or a bureaucrat in the Pentagon, you have to understand their perspective, their framework, to be able to interact with them.

The summer after junior year another government-sponsored, Sputnik-scare spin-off came along: a month-long tour of Boston area medical schools and research institutes to get physics, chemistry, and other physical science students interested in the biosciences. Professor Richter was my adviser, and since I was a chem major, he included me in the tour.

For me, it was another terrific experience. We were lectured to by people like James Oncley, a pioneer in protein chemistry who was uniquely involved in studying the plasma proteins and their properties. We saw electron microscopists studying nerve membranes and a whole variety of people applying advanced techniques from the physical sciences to the study of biological processes.

Boston was as far away from home as I'd ever been and a wonderful introduction to big-time science. By that time, I'd pretty much made up my mind that I wanted to be a doctor. Beyond that, though, I didn't have much of a clue.

Bill Foege, the distinguished and compassionate former director of the Centers for Disease Control, who went on to head the Carter Center, used to remark that the young, inexperienced EIS officers CDC customarily sent out to investigate mystery disease outbreaks and epidemics actually had some advantage over their more experienced and seasoned elders. While having first-rate training and the support of the entire CDC organization, they hadn't seen enough to have preset opinions and might therefore have been more open to new possibilities and had the energy to pursue them.

That pretty much sums up my approach to medical school. I applied to a couple of schools I thought I could get into, then several I'd heard by word of mouth were really good: the University of Texas at Dallas, Washington in St. Louis, Rochester, and Johns Hopkins. Much to my surprise, I got in to all of them.

I ended up choosing Hopkins because it met all my criteria: it had a good reputation, they let you know how much financial aid you could expect as soon as you were accepted rather than making you fill out another application, and it was located in a sophisticated eastern city and the only other place like that I'd been in before was Boston. Also, I had a friend at Hopkins who said how much he loved it there, unlike another friend at Harvard who kept telling me how much he hated it. So I packed my bags for Baltimore and didn't look back.

I'll never forget seeing the hospital for the first time. The building itself is a grand brick building with a breathtaking dome. Back in those days there was a sundial out front with the engraving: "One hour alone is in thy hands, the hour upon which the shadow stands."

That was intimidating enough, but there was an overwhelming sense of "those who came before." If I'd been more sophisticated, I might have focused on the fact that perhaps no other institution of medicine throughout the world had such an illustrious tradition or had been populated by such legends. Within this edifice walked the ghosts of William Welch, one of the founders of the hospital and the first dean of the medical school; Sir William Osler, author of the landmark textbook *The Principles and Practice of Medicine,* and perhaps the greatest physician philosopher; William Halsted, the great surgical innovator; and Harvey Cushing, the pioneer of neurosurgery.

The classic triad they preach at Hopkins, one of Osler's many legacies, is the equal importance of research, teaching, and patient care. You cannot do any one of these three well, they believe, without also being good at the other two.

This august sentiment was in sharp contrast to the rest of the physical environment there, specifically the high-rise dormitory in which I was to live. Set squarely in an urban ghetto environment for which my rural Texas upbringing had ill prepared me, the Medical Residence Hall was a box of smaller boxes which served as single rooms, with a common shower for each floor. The rooms were clean, but the theme was stark institutionalism—narrow, built-in bed, small closet, bookshelf, chair.

Outside, of course, things were worse. For someone like me from West Texas who was used to everyone having guns but no one using them, it was like a war zone. You didn't go out after dark if you could help it, and if you did, it wasn't for a moonlight stroll. You'd hightail it straight to the hospital or library or wherever else you were going. Several nurses on the late shift were raped during my first semester, one a couple of blocks from the hospital, just outside the window of a friend of mine as we unknowingly heard her screams. As poor as most of us were as medical students, we were still in a higher economic bracket than most of our neighbors, if for no other reason than that we knew our poverty was temporary. As I became exposed to desperation and deprivation throughout the world, I quickly came to realize that this is the largest issue in public health; the bugs themselves take second place. And urban Baltimore was my first exposure.

Hopkins Medical School worked on the quarter system. Instead of taking four major classes which ran all year, you had one each quarter, which turned out to be a good way of devoting yourself to each subject in depth without any competing priorities.

Though I didn't know much about him at the time, William Osler's influence permeated almost every aspect of the school. Describing the way a doctor ought to educate himself and the kind of commitment he felt a doctor should have, the great Canadian physician had written, "He does not see the pneumonia case in the amphitheater from the benches, but he follows it day by day, hour by hour; he has his time so arranged that he can follow it; he sees and studies similar cases and *the disease itself becomes his chief teacher,* and he knows its phases and variations as depicted in the living; he learns under skilled direction when to act and when to refrain."

This is certainly a lesson anyone practicing in the field of public health, faced with the outbreak of unknown diseases among a patient base you know little about, needs to understand and take to heart. It is clearly something that has stayed with me throughout my career. And I think by the time I got to the clinical years and was exposed to the institutionalization of Osler's philosophy, I'd already internalized it to a certain extent, already understood it instinctively. My mother and her illness were my first teachers.

Unlike many med schools where responsibility for teaching students is passed progressively down to the lowest guy in the pecking order, at Hopkins senior people gave the lectures and taught the labs. The head of the biochemistry department, Al Lehninger, used to sit in on all the lectures, taking notes as if he were a student. And knowing the head of the department would be listening intently probably didn't hurt any of the professors' preparations. Lehninger, by the way, went on to write what became the standard med school biochemistry text, so I'm sure he didn't feel his time at those lectures had gone to waste.

For most students, anatomy class is the first time you come in contact with a dead body. I'd worked at the hospital lab back in Texas, so I'd already started to develop some sense of clinical detachment about illness and death—separating a person's disease from his or her personality—but it was still a major adjustment to get used to working with a cadaver. Among my classmates there was a spectrum of reactions, from revulsion and nausea to outright horror and sick humor. I'd like to say most of us are too sensitive to indulge in the latter, passing off severed hands or penises to unsuspecting acquaintances and the like, but I'd be lying. In any business as serious as medicine, laughter is the most effective coping mechanism. As long as you never lose your respect for the dead or the great gift they've given us by willing their bodies for education, then you're far better off and you'll last a lot longer if you can maintain a sense of humor.

Still, you can never forget what you're dealing with. I remember a visit I made to Mexico City during a rare med school vacation after junior year. Through a Puerto Rican friend, I met a medical student there who took me deep into the basement of the medical school, down a long, darkened hallway, until we came to a huge room at the end. It was a giant cooler, and attached to the ceiling there was a pipe that ran the length of the room. All along the pipe was a row of hooks, and from each hook a corpse hung, suspended in the air, dangling by its ear. Outside in the hallway sat a group of little old ladies, burning candles and praying for the souls of those inside. I've seen a lot of gruesome things in more than thirty years in medicine, but that sight in the catacombs of the medical school in Mexico City is about the eeriest thing I've ever encountered.

Physiology came easier to me. The head of the department was a man named Philip Bard, and—this will come as no surprise—he authored the standard medical school physiology textbook. Although he was approaching retirement age (the age at which Osler had once gotten himself in trouble by humorously suggesting doctors ought to be put permanently to sleep), Bard was very hands-on and would come down in his white lab coat to interact with the students. His favorite lab subject was brain-lesioned cats, which was where we learned about neurological symptoms and diseases.

I also got my introduction to infectious disease epidemiology early on at Hopkins through Bob Fekety, an ex-EIS officer. Fekety gave us one of the classic EIS kinds of problems to work on, the kind with many more implications than the apparently simple problem at hand. The case was what we call "Eat/Don't Eat" epidemiology, in which some people in a group have a food-borne disease—or you suspect they've got a food-borne disease—and you make a list of all the foods that have been served and you go down the list asking the sick ones, "Did you eat this? Did you eat this?" Then you ask the same questions of the healthy ones in the group and you compare the two. We call it a case control study.

The case Fekety gave us was based on a true story of an epidemic in the officers' mess on an army base that was traced to the salad dressing. The guy who made the dressing was an enlisted man who had hepatitis and who was known to be notoriously hostile to officers. His hostility was reputed to have manifested itself in his urinating in various dishes. And so, although it was never conclusively proven, the belief was that he had urinated in the salad dressing and given all the guys who ate salad there hepatitis. Compared with more droll subjects like anatomy, epidemiology provided a terrific break in the routine and I think perhaps on a subconscious level Fekety had started the process of my conversion to a life of combating infectious disease.

It's in pathology that you begin to understand disease versus normal body processes. Up to this point, med students like me would hang around outside the emergency room, hoping some tired intern or resident would allow them to stitch up a cut. We'd have both hands and a brain full of lectures, slides, and guts from our deceased "patient" in the anatomy lab and we were desperate for a shot at making a difference—no matter how small—in the life of a real, live patient. My pathology section was taught by a guy named Abou Pollack, who could do things with an ordinary microscope and simple stained tissue sections that are hard to improve on today, even with all the scientific equipment at our disposal. He could walk you through the slides from a cadaver and make you see the entire disease process unfold before your eyes in absolute detail, with all the causative implications. Hopkins was really set apart—and ahead—as a teaching hospital by its rigorous program in pathology. They made an effort to obtain permission to do autopsies on all the patients who died in the hospital, which was critically important to the training program and to research into new diseases and the causes of old diseases. As the hantavirus outbreak in the Four Corners region underscored, you can't overstate the value of good postmortem information in discovering the true cause of death. For example, we'd spend hours looking around the thyroid and the thymus in excruciating detail, hunting for the last of the four tiny parathyroid glands.

It was common to have the physicians join us in the autopsy room, trying to learn anything that could help them save the next patient. Phil Tumulty was a professor of medicine and so respected that he was the one most of the Hopkins faculty went to when they were sick. He was at least sixty and had a national reputation. Yet he never missed an autopsy on any patient he'd been associated with, figuring he could always learn more. That was really part of the research/teaching/patient-care trinity at Hopkins, and it's disturbing today that despite their importance as a diagnostic, learning, and research tool, the number of autopsies continues to drop each year, primarily because there's no funding for it. Insurance doesn't cover it, next of kin rarely want to pay for it, and there's no extra money floating around to support it.

As I began my junior year, I also began another life journey of sorts: I got married. Lea Swinton was a technician who worked in the neuropathology lab. Blond, slender, and full of energy, she caught my interest as soon as I saw her. We were thrown together, in a sense, because we both lived and worked in the inhospitable surroundings there: the crime rate was high, personal assaults were high, deaths were high, and we found comfort in each other. We enjoyed each other's company, felt strongly emotionally attached, and got married; it was as simple as that. Despite the pressures of our environment and my studies, we had some genuinely good times. Over the summer, Hopkins would give us a free quarter to choose an elective. I elected to work with Stuart Tauber at Parkland Hospital in Dallas, doing research into calcium metabolism. We threw a few things in the car and drove down, which gave Lea an opportunity to meet my family. They all got along well, especially Lea and my dad.

On the research side, our experiments involved extracting the parathyroid hormone from the glands of cows the lab procured for us at a nearby slaughterhouse. We would go through a huge pile of bovine parts, snipping and removing the parathyroid glands and extracting the hormone in what had to be a very bizarre scene to any casual onlooker. The actual purification process involved precipitating the hormone in huge quantities of ether—the handling and disposal of which was not regulated then as well as it is today. For safety reasons, we would wait to use toxic and flammable chemicals until the weekend, when most people were away from the building. One such weekend, in our lab on the top floor, we were hard at work extracting parathyroid hormone and pouring used ether down the drain—not considering the basic scientific fact that ether floats on water. It went down the drain, and some guy on the floor below flipped a match into the sink and got the surprise of his life. Under great administrative pressure, we quickly developed more advanced methods for the disposal of ether.

That same summer, Lea and I took the trip to Mexico that, I believe, got the travel bug in me for good. (My only previous trip to Mexico had been a brief one in high school, in keeping with a West Texas tradition of teenaged males visiting a place known colloquially as "Boys' Town.") We got in our Ford Falcon and drove south to Mexico City, stopping along the way to sightsee and practice our high school Spanish. When we got to the ruins at Teotihuacán, I had an almost mystical experience. The pyramid there is huge and it's set out on the plains in one of the high mountain valleys in Mexico. It was an overcast day and there were almost no tourists. The clouds closed in and suddenly there was dramatic thunder and lightning. Standing there amidst the natural forces and imagining the great cultures that had been there before, it was tremendously moving and romantic. It was my first exposure to ruins, and since then the ancient and abandoned buildings of past civilizations have fascinated me. Whenever I am traveling to investigate an epidemic, I try to see the traces of the culture that preceded the peoples with whom I am dealing.

That trip was full of emotional contradictions for me. Lea and I shared the moving experience at the pyramid, then I visited the ca- daver storage room at the Mexico City medical school. We got to meet the famous microbiologist Raúl Castañeda, who took time out to take a lowly medical student from the States on rounds. Along the way, he revealed a political outlook far to the right of my own— which apparently Lea shared with him. There were times during that trip when, had I been a little more insightful, I might have seen potential trouble spots in my marriage, but we had so much fun eating and drinking in tiny, cheap local restaurants and visiting his- toric sites that I didn't pay attention to the things that might have bothered me. It was the kind of trip—and the kind of marriage—that you can make work only when you're very young.

Back in Baltimore, my first rotation was at Baltimore City Hospi- tal. I can't stress enough what a great training ground the old city and county hospital system was. When I was a student, the private- patient wards were regarded as an inferior place to do medicine by comparison because you had no responsibility for the patient. It was up to the private physician to order the tests and plan the treatment strategy and you were left just hanging around watching. But in a public hospital like Baltimore City Hospital, it was yours from the word "go." You had all these people to deal with, presenting a vast array of illnesses and symptoms, and you quickly learned to think on your feet. You also learned when to call for help. The local city and county hospitals are classically where the action's been: places like the Osler Wards at Hopkins, Boston City, Cook County, Bellevue. When they were well supported in terms of infrastructure and mate- rials, they attracted the best interns and residents with the best medi- cal school faculty looking over their shoulders.

Later, when I was doing my internship at Parkland, we would admit every third day and we'd have three to six patients who were all sick as hell keeping us up all night. On an easy night, you had to be on top of at least two or three of them all the time. But if you can handle that, you're in a lot better shape to handle other things in life. Today, I can often tell the difference in the attitudes and responses of people who haven't had this type of clinical training. When we get in tight situations at CDC—when the phone's ringing and it's a state epidemiologist chewing your ass on one line, a doc who thinks he has a hantavirus pulmonary syndrome patient on another line, the press on another line, and your boss is demanding to know what to tell the press—you've got to be able to balance it all.

When it came time to choose an internship, I wanted more direct contact with the actual faculty rather than time with the postdoctoral fellows, as had often been the case during my last two years at Hopkins. I was interested in Parkland because of their reputation of having young faculty who worked with the students and the residents. I would certainly enjoy being close to my father again, but I have to admit that it was more the proximity to senior people in the hospital that attracted me. I visited all the places on my list and gathered data from the med student grapevine. Although I was impressed after my visit to Boston City Hospital—the hard-core guys who were making the big medical contributions worked directly with the house staff—I wasn't interested in going someplace where I'd be doing even more scut work than I'd done at Hopkins. The guys at Boston City had to run their own blood tests and such themselves. It wasn't that I felt I was above doing those things, but I'd done all that as a high school senior at Ector County Hospital when I wasn't already putting in twelve-hour days. At Mass General, the folks were just too full of themselves and too busy to take the time to talk to me; I really had to twist some arms to get answers to my questions and I was put off by that. The esprit de corps was more in evidence when I went to visit Yale, which seemed a fine place to work and learn, but it just wasn't Parkland. What Parkland had above and beyond all these other places was a certain energy and drive and openness in the nature and character of the people who worked there. I'd been fortunate enough to be exposed to my father, Mrs. Willie Filleman, Zorus Colglazier, Philip Bard, and I wanted to continue to surround myself with people like them.

The same time that I was considering the different places for my internship and residency, the necessity of preparing for the even more distant future reared its ugly head. I was a senior medical student in 1966, and while it was too early for me (along with most of the rest of the country) to be overly concerned with whatever was going on in Vietnam, I knew once I finished my internship I'd be pulled into the military and be stuck being a GMO—a general medical officer (or "pecker checker" in the popular military medical parlance)—if I didn't have something else lined up. I'd talked to friends and I knew what happened to the guys who went to the army after internship: you spent about four weeks at some place where they taught you how to wear your uniform, which side your badge went on, how to march around the block, how to read your pay stub, and then they'd give you a couple of lectures on gunshot wounds that really didn't tell you much. Then they sent you, usually, to some place where your major duties were doing physicals on GIs at a rate of thirty an hour and clearing up gonorrhea when the boys came back from a night out on the town.

So I applied to the Public Health Service to go to the National Institutes of Health, or NIH, which was known as the place where all the best research was being done. I went over to the Wyman Park Public Health Service Hospital in Baltimore to get my physical, and the physician, sort of an older, slightly corpulent man, looked at me and laughed. "Oh, one of those Hopkins guys applying for the Public Health Service. Well, they're all hypertensive and they wash them out and put them in the army." I guess he'd seen enough stressed-out "Hopkins guys" to recognize one, because, sure enough, my blood pressure was high the first time he took it, but after I exercised, it came down. I applied and was accepted, so although I didn't know exactly which program at NIH I'd be a part of two years down the road, I knew somewhere there was a place for me once my residency at Parkland was finished.

When I finished medical school, Lea and I hitched a trailer up to the Falcon and drove down to Dallas. We rented a little house not far from the hospital and things blossomed professionally as they exploded personally. The bottom line is: med school marriages—even ones with principals who are more innately suited to one another than we were—rarely work for the long term. Back in Baltimore, which is where she grew up, Lea had a whole support system of friends to turn to while her husband busied himself in the lab. Although she got a job doing histopathology in the lab in Dallas, it was never really her place the way Baltimore had been. And on my part, I was way more stressed than I'd ever been in med school: the hours were longer and I enjoyed the work so much that I tended to disappear. At home, on those rare occasions when we were both there at the same time, we grew increasingly hostile to one another.

Although no one event marked the death of our marriage, there was an incident that illustrates the competing priorities that doomed the relationship. One night Lea and I had planned to go out for a romantic dinner. We already knew we were in trouble and this was meant to be a wonderful healing night for our marriage. I was getting ready to leave work when the lab called: "The test results are back on Mrs. Such–and–Such and it looks like she has some acidosis." So I went over, reviewed her lab results, and looked at her, and she really didn't look so bad. She didn't look great, but I didn't believe the lab results. I drew a repeat lab test, sent it off, and asked the guy who was supposed to cover for me that night to check in on her results when they came back. Then I went off for my night out with my wife.

The next morning I went in to find an empty bed, which is always a bad sign. The room looked like a battleground. My patient had gone into severe lactic acidosis and they were unable to save her. Now, in most places, it's perfectly acceptable to sign your patients over to someone else—it's standard practice. But at a place like Parkland, especially for someone coming from a place like Hopkins, the expectations are different. In those places, you don't leave a patient when there's any possibility that somebody who doesn't know her may get stuck with a major all-night struggle. Would my being there have saved her or otherwise made any difference in how things went that night? Probably not. But I vowed it wouldn't happen again to any of my patients. Win or lose, I'd be with them to the end. I didn't hold that incident against Lea because it had been my call to leave the hospital that night. But anytime after that when she called, urging me to come home when I just wasn't ready to go, a little resentment grew inside me—couldn't she see the stakes here? It wasn't too much later that Lea and I decided to throw in the towel. Although it wasn't a happy decision, it was mutual and I think we carried it off with decency. But even when it's the right decision, divorce is hard. We were already stressed at that point in our lives, and now our friends were choosing sides and all the rest of it. It's no wonder, then, that I just dug in deeper at the hospital.

Fortunately, there was a lot there for me. I genuinely enjoyed taking care of patients in a teaching hospital because when they come in they're completely unpredictable. They could have anything and often they're really sick, presenting a very complicated intellectual and human challenge. As I approached the end of my residency, then, in the back of my mind as I looked into the future, I vaguely knew I wanted to take care of hospitalized patients and I still wanted to do research, but I didn't know exactly how I'd do either. One thing I'd learned, however, was that my interest was starting to focus in the area of infectious disease. It wasn't a sexy specialty in the 1960s; ever since the large-scale use of antibiotics and stepped-up research in vaccines, people had been predicting the end of infectious diseases. Even in my naiveté, that seemed a bit premature.

One of my favorite subjects at Rice had been physical chemistry and at first I thought my research career would be spent applying some of the principles of thermodynamics to the transport of ions and water across biological membranes like kidney tubules. The nephrologists at Parkland were among the sharpest in the world and this helped steer me in that direction until I began to see more of their patients, particularly those that suffered from chronic renal failure. In the 1960s, there really wasn't much we could do for them. The patients I would see—many of them children—would have high waste levels in their blood because their kidneys weren't doing their job. They'd also be anemic, because kidneys make a hormone that stimulates bone marrow to make red blood cells. They'd smell of the waste that came out in their sweat. They might be obtunded (or fuzzy-minded), nauseated, maybe vomiting. They'd be itchy from all the toxins and waste materials coming through the skin instead of being excreted in their urine. Because of their anemia and fluid retention, they'd easily go into full pulmonary edema, sitting bolt upright in bed, severely short of breath. With some of them, you could even see the white frost on their skin from urea in their sweat. And they were just miserable. The only help we could offer them was dialysis, but since we didn't have the technology or the finances for chronic dialysis (and transplantation as a common practice was also way in the future), the hospital had a limit: you could only dialyze a patient three times. Now, if a patient had acute renal failure from some other cause, three times would be enough to keep him alive until his kidneys came back. But for most people, the three times provided just a brief respite before death. If it was the first time they presented with renal failure, they had some residual function left. Sometimes they'd go three or four months before they'd get in trouble, but sometimes they'd be back in a couple of weeks. And I had to explain to their family—their spouses, children, parents, lovers—that this was it: "We can only do it three times and that's all."

"Well, won't you try it again?"

"Sorry, we can't do it; it's too much for their organs." I found it just too debilitating to treat people with such a temporary solution, knowing I really couldn't help them.

Infectious disease, on the other hand, offered the promise of making a long-term, positive difference in patients' lives. At Parkland, the guy in charge of infectious diseases was Jay Sanford, who's still one of the best docs I've ever met. He was one of the main consultants in infectious diseases in Vietnam and was always flying around the world to advise—a highly talented field guy as well as a skilled, technically adept academic. Even before I met Jay, during my years at Hopkins, I'd used one of my free quarters to work on an experiment designed to offer physicians treating cancer patients with chemotherapy insight into how far they could go before a serious risk of opportunistic infection developed. We experimented on rabbits, giving them nitrogen mustard to make them neutropenic—destroying their bone marrow and decreasing their white blood cell count. Then we found out what critical level of white blood cells corresponded to an increase in susceptibility to bacterial infections, which is what was happening to many of those receiving chemotherapy. Today we know exactly what levels to look for, but back then this was research that would yield valuable results.

The staff at Parkland was incredible. The head of medicine, Don Seldin, had a real eye for recruiting good people. The environment was one in which it was expected that you would work with the medical students and house staff and teach them as much as you knew at the same time that you were taking care of patients and doing research. In a sense, this was the end of an era in medicine. I don't know whether today you can keep those ideals active in your career and still expect to have any kind of life. Budgets are too strapped, human resources overstretched, people pulled in too many directions, science too complicated.

I'm afraid what's been said is true: that doctors of my generation and the generation who taught us are the first for whom medicine was a science and the last for whom it was an art.

3

MARU

IN THEIR most idealistic and romantic moments, most young doctors imagine themselves as single-combat warriors against death. And, in fact, the warrior analogy was not lost on the powers that be during our country's development. The earliest version of our Public Health Service, the Marine Hospital Service, had initially been established to run seamen's hospitals, but under the onslaught of a deadly disease of the late 1800s, yellow fever, it had developed a cadre of uniformed officers who ran a string of hospitals and quarantine stations across the country, and investigated infectious diseases such as yellow fever, cholera, and smallpox. This organization became the Public Health and Marine Hospital Service in 1902, and later the U.S. Public Health Service, continuing its tradition of maintaining a group of professionals commissioned with military rank who had the expertise to marshal resources to fight disease on a national level.

I was accepted into the Public Health Service during my senior year at Hopkins, with the idea that I would begin after two years of internal medicine training. The PHS had several programs I could choose from: one headed by Bob Chanock that involved studying respiratory and enteric viral diseases as they broke out in schools and communities; the other a position working with Karl Johnson at NIH's Middle America Research Unit (MARU) in the Panama Canal Zone studying Bolivian hemorrhagic fever and other tropical viruses. Both looked pretty interesting.

I saw this as a choice where neither was obviously better, neither would close any doors, and I could learn a lot from either. But I remembered how much I enjoyed visiting Mexico with Lea. We had gone back several times bumming around on a shoestring budget, drinking with the locals, taking in the sights, riding the buses, and staying in romantically crummy hotels. I figured, hell, I've been through four high-pressure years of college, four years of higher-pressure medical school, and two years of internship and residency where I had almost no time to do anything, so I decided to go to Panama: join the Public Health Service and see the world!

Since I was given the choice of how to get there, I packed up my car with all of my significant worldly belongings and drove. The drive itself was beautiful, and as I passed through Mexico and down into Central America, I grew increasingly glad I had picked Panama for my assignment.

I arrived on a Saturday and couldn't get anyone on my xeroxed list of people at MARU on the phone. Tired from my journey, I checked in for the night at a place called the Hotel Ideal, so named by a proprietor who believed he should offer a cheap, clean place to stay where the hungry and tired could also get inexpensive (if plain), nutritious food to eat. It was ideal for me since it was cheap and they even had a place where I could put my car and it wouldn't get ripped off.

The next morning, Sunday, I reached Karl Johnson by phone at home and he gave me directions to the house in the Canal Zone where he and his wife Patricia Webb lived. While the directions he gave were easy to follow, I had some trouble making the necessary mental and emotional adjustments as I crossed over into the Zone.

The Canal Zone was spread along each side of the Panama Canal and with the way the Central American landmass curves around at that point, it ran more east and west than north and south. Everything in the Canal Zone was built for the American establishment there. Houses like the one Karl and Patricia lived in were the sort of open tropical designs you might find on a military base anywhere in a hot climate. There were several bases with small dispensaries as well as Gorgas Hospital to tend to American military and civilians. Gorgas—named for William Crawford Gorgas, one of the military pioneers of yellow fever and malaria control—was an older facility with a modern, air-conditioned new building attached. The older buildings were the type of hospital the U.S. military might build anywhere in the world. Not designed for the tropics, they lacked air conditioning and had windows so small it looked as if the architect had been preparing for a siege. The Social Security Hospital in Panama, built by and for Panamanians outside the Zone, also had no air conditioning, but at least it sat atop a hill. Its big, screened windows caught the breezes which carried across the long corridors, making it a much more comfortable facility to visit for both patients and doctors.

For three weeks I'd been driving across Central America on roads where, regardless of the country, all citizens drove as fast as their cars could go and where no speed limits were posted. A stop sign meant that if you drove into the intersection and hit somebody, you should have stopped. At intersections, you had to demonstrate your machismo and intent to proceed or those positioned to cross your path would assume it was their turn. You could show intent by a hard look, by not slowing down, blinking your lights, honking your horn, or any other unofficial signal that seemed to fit in with the Latin culture.

Traffic lights worked the same way. They were uncommon, and where they existed, they were more decorative than functional. As in the world of the microbes, it was survival of the fittest—yield in any way and you've lost your claim to that territory.

As I left the lush and beautiful Panamanian countryside and entered a neighborhood of cookie-cutter, American military-style houses, I could sense that a different mentality governed here, but in my mind, I'd already gone native. As I drove up the street, I saw Karl's house on the left, so I pulled over and parked in front, facing the wrong way.

There was zero traffic and it was Sunday. But the first thing Karl said as he greeted me was, "You're going to be in trouble. You're parked on the wrong side of the street."

I appreciated his concern but reassured him. "Well, you know, I just spent three weeks driving through Central America and it hasn't been a problem."

"Yeah," Karl said, "but you're in the Canal Zone now." It wasn't half an hour before we were interrupted by the appearance of a Canal Zone policeman in his knife-edge creases and his Smokey the

Bear hat demanding that I turn the car around. I learned fast that to survive in Panama you had to understand and work with the differences between the two cultures wherever you encountered them.

Karl Johnson and Patricia Webb were, in some ways, the "Hollywood" couple of public health. Karl was a dynamic young doctor who originally had come to NIH—the National Institutes of Health in Bethesda, Maryland—to work with Bob Chanock on his studies of respiratory viruses in children.

Patricia was a medical virologist who'd worked overseas with rickettsiae, small bacteria which have a number of properties in common with viruses. And when she came to NIH, they fell deeply in love. The only problem was that they were both married with families. Karl had three children and a loving wife who stayed home with them. Patricia had two kids of her own and was married to a distinguished rickettsiologist named Bennett Elisberg.

The two of them became quite an item and later were married. It was rumored that the brewing scandal was one of the reasons Karl went to Panama. I'm not completely sure why, but the fields of epidemiology and public health seem to promote as many multiple marriages and divorces as the film industry. Maybe it's the pressure of the work, or the temperament and passion of the individuals who go into these fields.

Not long after Karl arrived in Panama, he went to Bolivia to investigate an outbreak of Bolivian hemorrhagic fever and caught the virus himself. This is a life-threatening disease, and Karl was brought back to Panama for treatment. Patricia came down from Bethesda to nurse him back to health. She then returned to Bethesda, at which point she got sick; she'd caught BHF from Karl. Eventually, after she recovered, she moved down to Panama to work with him.

As I sat across from Karl and Patricia following introductions and small talk, he said, "Well, I'd like you to work with Patricia on a new arenavirus that we've found in Bolivia." This was a cousin of Machupo, the Bolivian virus that had almost killed Karl. But this one was not thought to be so highly pathogenic.

The lab at MARU handled some of the deadliest and most infectious diseases known to medicine at the time (Ebola and AIDS wouldn't be discovered for a decade or two), long before sophisticated equipment, protocols, and procedures had been developed to handle such biohazardous agents.

The laboratory was set in a white, generic military-style building up on top of Ancon Hill in an area known as Ancon Heights. Originally used as nurses' quarters, it was now divided between a U.S. Army parasitology unit and the Public Health Service, which primarily conducted virology research. There were administrative offices on half of the ground floor (shared with the army), labs on the second and the third floors, with the fourth floor left as semifurnished living quarters for visiting guests: scientists from countries ill equipped to provide them with a per diem, for example.

Although we didn't have the extensive biosafety level setup available today, Karl had arranged things to provide the greatest level of protection possible for both the workers in the building and the community outside. Karl's system relied heavily on the work done at the biological warfare labs at Fort Detrick, Maryland, in the 1950s. Different rooms, physical barriers, and directed air flow were used to separate potentially deadly agents from susceptible populations. For example, the labs on the second floor of the building dealt with many different viruses, but work with the most dangerous and infectious diseases was restricted to the labs on the third floor. Half of that floor was blocked off for Machupo, the virus which causes Bolivian hemorrhagic fever. This disease had almost killed Karl and is highly aerosol-infectious. Machupo is so infectious and deadly that the Russians had a vigorous and chilling interest in it during the Cold War, and were suspected of trying to make it into a biological warfare agent. They went to Bolivia to get their own strains and suffered several laboratory disasters when they brought it back.

So the MARU lab was set up with containment in mind, and Karl and his people understood well a rule that is just as true today as it was then: equipment alone does not constitute a safe environment. You have to have well-established procedures for keeping everyone safe (inside and out) as well as appropriately trained personnel. The people who went into the labs to work with infectious agents all wore scrub suits and entered the lab through an air lock. When they came out, they had a place to take off their suits and shower before getting dressed to rejoin the outside world. There were big blowers set up so that air pressure in the lab was always negative to the hall, meaning that if a virus became airborne, it was sucked back into the lab. The air exhausted from the labs passed through HEPA—high-efficiency particle—filters, so people walking around outside the building had no chance of breathing in virus particles. There were also redundant blowers and power supply, so if we lost power there was backup.

But the greatest fail-safe mechanism was the people. All well trained and disciplined, the only people allowed to work with Machupo virus were also those who, like Karl and Patricia, had already had Bolivian hemorrhagic fever, since once you had the disease and recovered, you were immune to the virus—much as you are with measles. The technicians hired to support the scientists at the lab were all carefully recruited from the Beni region of Bolivia, where the disease is endemic, were known to have had it, and had tested positive for antibodies to the virus measured by the only test available at the time, the complement fixation (CF) test. They had no formal education in medicine, indeed many had never finished high school, but they were all bright and well trained. If they didn't know the medical or scientific word for something, they'd make it up. We'd take out a spleen sample and see *espleen* written on the slide. Or if we were doing research with testes samples, the slide would often be marked *cojones,* rather than *testiculos,* since that was the word they knew for this particular part of the male anatomy.

As enlightened as Karl's lab procedures were, there is often a gap in science between theory and reality. If this weren't so, everything we can make happen in the lab we should also be able to make happen in real life, and we know that just ain't the case. One of the Machupo lab technicians, a twenty-something fellow named Einar, developed a fever one day and stayed home from work. Within a week, he'd developed full-blown Bolivian hemorrhagic fever. A few days after that, he died. Karl was stunned. Einar came from the endemic area, he had a history of the disease, he tested positive for antibodies to the virus . . . and yet he still died from it. Karl was able to isolate the virus from Einar's tissues, so he was forced to conclude that Einar probably had not had the disease before. If we'd had a more specific test for antibodies, such as the virus neutralization test that was developed later, it would have proven he was not naturally immune to the virus, though he'd been tested and retested by the standards of the day before he was allowed in the lab and he had met all the criteria. It's possible he had had another disease in the arenavirus family similar enough so that the CF test couldn't differentiate it from Machupo.

It deeply affected everyone on the project and Karl had to fly back to Bethesda to defend the whole program to PHS brass. The experience also had a tremendous impact on Einar's people. I was in the endemic zone some twenty-five years later and met people who lived in the town Einar came from. They all still remembered the U.S.

Army DC-3 flying back with the sealed metal casket and officers delivering it and a citation from the U.S. government to Einar's family, along with money for burial expenses. I didn't hear anything but pride from them in what Einar had done for his country. Second-guessing ("Why did this happen?") came only from the people at the lab, who mourned his loss and dedicated themselves even more to unlocking the secret to the virus. I came right after Einar went. I knew him as a screw-capped vial called "Einar—brain."

Before I was able to get settled in my work at the MARU lab, I had an opportunity to test out the medical facilities at Gorgas Hospital firsthand. In my drive through Central America, I had eaten the native dishes even though I hadn't taken my gamma globulin before my hurried departure. I figured I'd be safe as long as I didn't drink the water or use ice.

When I started feeling constantly tired, depressed, sleepless, and feverish at night, I couldn't help but think of one particularly delicious but questionable dinner on the border crossing from Mexico to Guatemala, where I drank beer from a glass that was washed in a little polluted stream that ran by the dirt-floored "restaurant." I suspected I had either hepatitis or tuberculosis, and went to see Gabriel Kourany, a hematologist at Gorgas whose sister-in-law worked at MARU. He ordered a blood count, being a hematologist, and when it came back normal, proclaimed me healthy. More depressed than before, I bought a bottle of port (a wine I haven't been able to drink since), drowned my sorrows, and went to bed. I woke up in the middle of the night feeling sicker than before.

Now, however, I had proof of my illness: the whites of my eyes were yellow, my urine was dark brown, and my bowel movement was white—clearly my liver wasn't functioning properly in excreting waste products into my bile. I knew then I had hepatitis. As I waited for morning, I alternated between relief that I finally knew what was wrong and fear that it was really serious. Although I had a much less extreme case than many of the people I'd treated back in Texas, I knew that some people die from hepatitis. I replayed in my mind like a mantra "many are infected, few perish," and as soon as the sun came up I called Gabriel.

"I've got to talk to you," I said urgently. "I have very dark urine."

He said, "Very dark? The color of Coca-Cola?"

"Darker!" I said.

I could hear the disbelief in his voice when he answered, "Okay, I'll meet you in the emergency room." Needless to say, after collecting a very impressive urine sample and running some liver tests, Gabriel admitted me to Gorgas, to one of the older, unair-conditioned wards.

There was an interesting ecosystem thriving at the hospital, developed as a function of economic, social, and military concerns. As a PHS doc, I was the equivalent of a military officer (we had uniforms, although we never wore them), which meant I was a step above any enlisted patient on the ward. The military caste system was so strong that when I was admitted and all of the four or five officers' rooms were full, they treated me in a windowless, converted linen closet rather than risk my fraternizing with the troops on the ward. The closet was so hot I begged the staff, as I puked and stained the sheets with yellow bilious sweat, to put me out on the wards.

They refused because I was an officer. "But there's a guy in here with lung cancer, and when he dies you can have his room. Don't worry. It won't be long." Better I should die of heatstroke than interfere with the natural order of things.

Soon, however, I was moved into a double room, which I shared with a short, fat, loud guy who was fairly high up in the Panama Canal Company. He snored like thunder over the plains all night and watched the Armed Forces Radio and Television Service (affectionately referred to as "A-Farts") all day—guaranteeing that I got sleep only during the two or three hours a day that he was taken away for physical therapy. Soon the fellow with cancer died, though, and I was upgraded to my own officer's room.

Once I was discharged from Gorgas, Karl kept me busy with research, including a stint involving a unique antigen found in hepatitis patients—a group now a bit closer to my own heart and soul. Socially, however, my life was still in the doldrums. I was not exactly viewed as a catch by the parents of Panamanian women, and there were relatively few single American women. I was not unhappy, then, when a new opportunity for field research presented itself in the form of an outbreak of Venezuelan equine encephalitis (VEE), which was decimating horses and burros in Guatemala. Discovered in the 1940s, VEE breaks out in very abrupt, aggressive epidemics that are lethal to horses and mules and frequently incapacitate humans. Then it disappears, only to return some unpredictable number of years later. Because of the important economic role horses and burros played in Latin America, beginning in the 1950s they were often vaccinated against VEE with crudely prepared, inactivated vaccines. In the late 1960s, however, the disease demonstrated a new wrinkle which brought it to the attention of MARU. It began to move from an epicenter in Guatemala into Mexico and other parts of Central America. If it reached "El Norte" it could wreak havoc with the large and lucrative ranches in Texas—and threaten U.S. citizens.

Even if it were not a threat to the United States, there were many reasons why VEE demanded our attention. Simply from an economic standpoint, horses and burros are vital in Central America. In Latin America the disease is known as *peste loco* because it affects the animal's brain. In the beginning, the horses act a little funny. They go off their feed and run a fever. But once the virus gets to high levels in the brain and the horses begin to suffer encephalitis, they become irritable, are difficult to manage, run around in circles, and eventually lie down, thrashing the earth in convulsions, and die. The mortality rate is very high; at one point, you could walk through sections of Costa Rica and there'd be dead horses everywhere. That was bad enough. But since VEE is spread by mosquitoes—which bite humans as well as horses—and since humans are also susceptible to the disease, it was pretty easy for the virus to cross over to human populations. Although it has a much lower mortality rate than it does in horses, it's a terrible disease. Most people who get VEE develop an acute fever, described as flu-like, although these patients don't have the cough associated with the flu. The disease has an incredibly abrupt onset—sometimes from one minute to the next—of high fever, chills, severe muscle pains, prostration, headache, pain behind the eyes, and intense photophobia, or aversion to light. You go in to check on one of these patients and you'll find them off in the darkest corner of their hut with the bedclothes pulled up over their face, lying there moaning, unable to move. The vast majority who get this reaction are totally incapacitated for three or four days before ultimately improving. There is no treatment, but within a week or two most of those who come down with the virus fully recover. However, about 1 percent or so get encephalitis (technically, inflammation of the brain)—anywhere from mild to severe—with any or all of its symptoms. We felt if we had a chance to make a difference in containing an epidemic, we absolutely should try to do so.

Nobler designs aside, however, the U.S. government had other reasons to be interested in VEE. The symptoms in humans are so incapacitating that VEE had been seen as a potential biological weapon. The army wanted to develop different categories of biological warfare agents: incapacitators as well as killers. With a relatively short incubation period of two to three days, VEE could be an ideal incapacitator: neutralizing an enemy population right before a battle without risk of killing innocent civilians or committing wartime atrocities. With that as a plan, the army had developed a vaccine to protect our troops in case an enemy tried to use it on them, or presumably in case the wind blew the wrong way the day they tried to use it on someone else. In the Latin American situation, we needed to test the vaccine in horses. Karl believed that only an effective, fast-acting, live-attenuated equine vaccine could stop the spread of VEE in horses. By immunizing the horses we could prevent infection of the mosquitoes that spread the virus to humans. The inactivated horse vaccine in use before had clearly not prevented epidemics.

There are two types of vaccines: killed (which uses dead viral material) and live-attenuated (which uses a much smaller amount of "weakened," living virus). The Salk vaccine for polio is an example of a killed vaccine, whereas the Sabin polio vaccine uses a live-attenuated approach. Each type has its own risks and benefits. The benefit of a killed vaccine is that the body will develop an immune response to the virus despite the fact that the viral material introduced is, in fact, dead. The risk is that quality control during production has to be perfect, to ensure that the virus really is 100 percent dead, or someone receiving the vaccine might develop the disease. In fact, in one of the army's first attempts at a killed VEE vaccine, the virus was incompletely inactivated and actually gave some people VEE in test situations. You run a similar risk with a live-attenuated vaccine. Typically, live-attenuated vaccine developers try to promote mutations that will lessen the ability of the virus to induce disease by growing the virus through many generations in cells cultured outside the body. The advantage over a killed vaccine is that a much smaller amount of viral material can be injected and allowed to multiply in the body and trigger an immunological response, which makes it cheaper and easier to produce. Live-attenuated viral vaccines are also known for inducing solid, long-lasting immunity. Most viral vaccines in use today (for example, measles, mumps, rubella, chicken pox, yellow fever) are live-attenuated. The drawback is that the mutations might revert—or go back to the original form of the virus—once the material has been injected in the body, causing recipients to develop the disease or, depending on the virus' method of transmission, maybe become infectious to those around them. A very small number of people (about one in 7.8 million) who receive the Sabin vaccine develop polio. One in 5.5 million passes along a reverted vaccine virus that results in clinical polio in a contact.

In many respects, this latest VEE outbreak epitomized the many problems inherent in the vaccine development process. Since we can't often rely on drugs to battle viral infections the way we can with antibiotics against bacterial infections, the best weapon we have against some viruses is a vaccine. Bacteria are easier to fight with drugs because they're independent, living organisms—they don't need a host cell to survive and they are capable of carrying out their own life processes. With a bacterial infection, therefore, we can introduce an antibiotic (literally, "against life") that works by inhibiting some of the bacteria's unique metabolic processes. With a virus (much smaller and simpler), we can't target its life processes because it doesn't have any of its own—everything it does is tied in to the host cell's mechanics. It does have some molecules of its own that give it an identity to target, but we can't combat a virus nearly as easily as a bacterium because it's too much like us—too much a part of us. So we must often turn to the body's own immune system to fight it, which is what vaccines are designed to do: get the body's own immune system to develop antibodies to a virus by injecting some of a virus vaccine.

Vaccine production is a risky, high-stakes gamble, almost as much art as science. The pluses and minuses of the two types have been argued for decades, a classic scientific debate. And all of us involved with vaccines are haunted by the specter of the infamous 1955 incident in which incompletely inactivated Salk vaccine resulted in 204 cases of poliomyelitis.

The problem is that with viruses spread by arthropods, the way VEE is, the only effective way to control them is usually through vaccination. What are these arthropods? Most people think of insects as anything that scurries around or buzzes and that you squash if it gets too close—in other words, all "bugs." But "insect" refers specifically to creatures like mosquitoes and sand flies with six legs. And many disease transmitters such as ticks, fleas, and mites have more than six legs. We use the term "arthropod" (segmented legs) to refer to the whole group.

You might think that aerial spraying would be enough to kill all the mosquitoes and stop transmission of the disease. But this kills only a portion of the adult mosquitoes, you have to deal with the environmental effects, and eventually resistance emerges. Thus, to eliminate the problem, you have to destroy the breeding sites—where they lay eggs—to be effective. And this is difficult or impossible to accomplish.

The bottom line was that the army agreed to let us test its live-attenuated VEE vaccine in the field. It was an opportunity we couldn't pass up, but we knew we needed to research the characteristics of the outbreak to really give it an effective field test.

Although the epicenter of the epidemic was in Guatemala, it was beginning to show signs of spreading into Costa Rica. Karl was sure that VEE was going to move right down Costa Rica's Pacific coast. He wanted a field team sent out there before the epidemic hit to bleed all the horses, collect samples to confirm which horses were susceptible to the virus, and be ready for the epidemic, a little like Babe Ruth pointing at the center-field bleachers before his home run. When the epidemic came, it would be possible to study the movement of the virus and also vaccinate some of the horses in a controlled study to test whether the vaccine worked. Karl deserves a tremendous amount of credit for both his foresight and his *cojones*. Few scientists would have been brave enough to put their resources behind such a risky prediction. Most of the time, epidemiologists find themselves in the position of cleanup crew. Like detectives called to a crime scene, they get there too late—the damage has been done, the body's on the floor. They can't study the early, revealing phases of the epidemic. But this time—if things went as Karl predicted—we'd be there early enough to spot the right clues, predict which animals were vulnerable, and maybe prevent the epidemic from running its course.

The first step was to get our people out there to bleed the horses fast. Two new guys were joining the lab then, Dave Martin (fresh out of Harvard Medical School, whose favorite phrase was soon to be "This is one of the things they didn't teach me at Harvard") and Bill Reeves (also relatively new to the field, whose manic energy and enthusiasm made up for his lack of experience). Karl sent them out immediately to bleed horses and people and pick up any other samples they could so we could test them. Actually, they ended up having to go out twice, because the first time they bled only the broken, "domesticated" horses brought by the ranchers at their request, which did not give us a representative sample. Much to their newly arrived wives' chagrin, Karl sent Dave and Bill back out to fill in the sample with the young horses, the old horses, the wilder horses out in the fields.

But just a few months later we hit pay dirt: in a scene reminiscent of *The Godfather,* the locals had cut the head off a horse dying from encephalitis and sent it on ice via DC-3 so we could test the brain. Peter Franck, who had taught me one of the basic tenets of field research—never go anywhere without a bottle of rum and a local map—excitedly gave me the news: "We've got VEE out of a horse brain from Costa Rica!" We had a confirmed case of VEE in the exact region of Costa Rica where Karl had sent our team to bleed horses. It was time to send in the troops.

We decided to send teams to Costa Rica in stages: I was part of the advance team with Gerry Eddy, an army veterinarian with a Ph.D. in microbiology, who'd previously been posted at the U.S. embassy as an adviser on animal diseases to the Panamanian government. Bill and Dave would follow once we'd set up camp. They had first rights to the epidemic, but they had just returned to Panama and their wives would have scalped them, and Karl as well, if they had just picked up clean socks and headed out immediately, particularly given David's recently born daughter. I have to confess that my outward professional demeanor did not match the little-kid excitement I felt at going out on my first full-fledged field assignment. The fact that this was also Gerry's first assignment did not frighten me in the least—in fact, it was a great opportunity for us both to learn as we went, unfettered by preconceived notions (or experience).

Our days-long drive to Costa Rica was breathtaking, particularly for someone from flat, dry West Texas. Through Panama, we drove by mountains covered with palm trees into cattle country with lush green pastures. At the border between Panama and Costa Rica, we reached a point where the road goes up a mountain separating the two countries, known as Cerro de Muerte, the Hill of Death, named for the Indians who died trying to cross it, unaware of the severe climate change from the bottom of the mountain to the top. Driving further, you see how easily one could make that mistake. You start out in the tropics, through rain forest, and gradually the flora is less and less tropical. As you move toward 12,000 feet, you pass hillsides full of wild hydrangeas in bloom, then, up where it starts getting colder and there's less oxygen, you start seeing lichens and mosses up past the timberline. And then it's the whole thing in reverse as you unwind back down the mountain to find yourself on the escarpment that runs down to the beach along the Pacific Ocean.

With this scenic drive unfolding before our eyes, Gerry and I were in a terrific mood by the time we reached San José. In Central America, climate is altitude. If you're at sea level it's tropical and it gets cooler and cooler as you go up higher. San José is up high enough so that it has lovely spring-type weather.

Our first stop was San José, the capital, because, typically, the first thing you want to do is check in with the local agency in charge and finalize your action plan. With a disease like VEE, it was especially important because the old turf wars were often going on between the animal health and the human health agencies. We dared not risk losing the cooperation of either by not giving each their due. Our main contact was the Minister of Agriculture.

The Ministry of Agriculture was typical of many Latin American ministerial buildings: turn-of-the-century stone, classically proportioned and enormous, with huge windows and high ceilings. With all the classic beauty, however, I instinctively felt that nothing had been modernized. The electricity was terrible, and although the building housed an important government agency, there were no copying machines or other forms of technology in evidence. As if to make up for this, though, our access to decision makers was unbelievable. In the United States, a field epidemiologist would never consult directly with the Secretary of Agriculture; he or she would report to someone way down on the food chain who would selectively pass information up to the next level, and so on. In Costa Rica, the chain of command was: Minister of Agriculture, then Deputy Ministers for Crops and for Animal Health. We met with the Deputy Minister for Animal Health.

We'd been on the road and I was suffering from traveler's diarrhea—almost the rule for fieldwork, as public health officers are subject to the same physical limitations as everybody else. At a break in the meeting, I asked a veterinarian on staff where the toilet was.

"Ah, yes," he said, "the toilet is down the hall. I'll get you the key." He handed me the key and added, almost as an afterthought, "Wait, I must get toilet paper for you, right?" I confirmed that would be most appreciated.

He conferred with the secretary, who went in the minister's office, took a key from his desk drawer, returned to unlock a drawer in her desk, and removed a roll of toilet paper. The veterinarian gave me the whole roll and I realized I was truly an honored guest. I graciously took my prize and used the key to unlock the ministerial toilet. There was a back issue of the *San José Times* perched on a coat hanger stuck over the door—presumably for the convenience of less distinguished users. The juxtaposition of elegant granite floors with the ratty old coat hanger came to symbolize much of my travels, where rich history and culture came with few modern conveniences. I tore off a length of tissue, stuffed it in my pocket, and mentally added to the inventory from Peter Franck: map, bottle of rum, plus toilet paper.

I returned, and the minister announced a new member of the group would join us, Dr. Walter Askew with the American embassy. Askew had come to facilitate the meeting and help us with a plan for controlling the epidemic. Upon hearing his name, Gerry turned to me, saying, "This is great. I know Walter and he's a good guy. I've known him off and on for years." Gerry and I were both optimistic, figuring that Askew would help us get our plan underway. That plan involved surveying the epidemic, tracking its progress, and trying to vaccinate in an area where the epidemic was headed so we could prove that the vaccine from the U.S. Army worked in the field. The vaccine had been tested in the lab extensively on smaller animals like rodents, and even on a few burros they kept in large containment suites, but not in the "real world." Our plan had the potential to stop the spread of the disease and provide valuable data on the vaccine. This might sound simple, but just as viruses are species-specific, plans to eradicate disease have to be virus-specific: you can't fight every epidemic the same way. We hoped the Costa Rican authorities would buy into our plan for combating VEE, especially since we'd had trouble getting our own government—the U.S. Department of Agriculture—to grasp the subtleties involved a year earlier. A meeting Karl attended with the USDA to discuss VEE control in the United States was a failure. The USDA had a monumental success with a very important disease of cattle and other cloven-hoofed animals, foot-and-mouth disease (FMD); they eliminated it from the United States in 1929 and have prevented its reintroduction since. This virus, like HIV, infects for the life of the host; FMD can reactivate and spread to new cattle as long as the original animal is around. To get rid of it, the USDA tested and identified animals who harbored the disease and destroyed them all, along with all the animals around them that could have been in early stages of infection when the virus wouldn't yet show up in tests.

That was an excellent strategy for foot-and-mouth disease. The catch is that the same strategy would be an abysmal failure with VEE because it is not chronic or latent and is transmitted by mosquitoes. Horses (and people) don't get the disease from other horses. They get it by being bitten by a mosquito carrying the virus. If you waved your magic wand and eliminated all the horses, burros, and people with VEE in Costa Rica, there would still be mosquitoes floating around with the virus, ready to infect the first horse you brought in from another country. Karl got terribly depressed when he learned that the

USDA planned to fight VEE in the United States the same way they did foot-and-mouth, killing all the exposed horses. Not only would this not prevent future outbreaks, but it would be destroying the most valuable horses, from an epidemiological standpoint, since horses that have had the virus and recovered don't harbor the virus and can't get it a second time. Once the horse makes an antibody response to the virus, it's immune—even if it gets bitten by another infected mosquito. Those are the animals you don't want to destroy. At the early stages of the 1969 outbreak, however, the USDA wasn't even prepared to do the wrong thing, preferring instead to watch and wait, as the threat moved aggressively into Mexico and closer to the United States.

Gerry and I listened as the Costa Rican minister described their plan, which involved vaccinating horses in the middle of the epidemic. It would do no good. VEE is such a fast-moving disease that it would be gone by the time they reached an area to vaccinate. The living animals they would be vaccinating would not be disease targets. They'd also screw up our studies. The disease would continue to spread and we would be unable to tell what the pattern was or whether the vaccine worked.

As the minister described this plan, Gerry leaned across Dr. Askew and whispered to me, "What asshole dreamed up this plan?"

As if on cue, the minister, who didn't hear Gerry's question, said, "We would like to thank Dr. Askew from the American embassy for helping us devise this plan to stop Venezuelan equine encephalitis in Costa Rica."

If there was a graceful way out of the room or the world, Gerry and I didn't know what it was. We bumbled around a discussion with our relatively poor Spanish and tried to interject the notion of maybe vaccinating ahead of things, trying not to appear to be discrediting Dr. Askew's plan before the minister. It was awful. The minister considered everything we said and then invited us to return the next day to finalize the plan. In retrospect, now that I've been through this routine many times, I can sympathize with the minister's situation. Here's one gringo advising him to do X and another gringo advising him to do Y.

He says, simply, "Well, ah, we should meet in the morning and discuss this some more."

For his part, Walter just cleared his throat and said, "Excuse me, I've got this cocktail party." Bonding among the gringos was over.

It was an incredibly depressing way to end what had started as such a beautiful, promising day. A couple of the veterinary and ministerial underlings invited us out to dinner and we went with them, wondering how we could have screwed things up so badly, so quickly.

The next morning, though, the deputy minister announced that after discussion with Dr. Askew, the minister made some tentative changes to the plan they'd proposed yesterday which, in effect, rendered it the plan we'd proposed. We immediately started preparing for our journey into more rural areas of Costa Rica. To support our combined effort, the cash-strapped ministry was going to take funds, human resources, and valuable vehicles and equipment away from other important projects. It took a day or two for them to get everything ready—time I spent rounding up several bottles of scotch at the embassy to use as goodwill offerings. If you're traveling on orders, you can usually go to the American embassy and get some liquor without any trouble, free of the very high local duty on whiskey there. Because of the duty, only the wealthy could afford such things. A fifteen-dollar bottle of scotch for us becomes a fifty-dollar gift, so we could buy cheap and give dearly.

When we got all the vehicles ready, we left for Guanacaste province, a day's drive from San José and located in the area we wanted to study. Bill Reeves and Dave Martin had joined our party driving a big, black International Carry-All, which was larger than our vehicle but short a muffler. We christened the loud vehicle the "Black Mother."

The motel in Guanacaste, El Bramadero, was fairly modern, 1960s decor with all the basics: bed with a broken mattress spring but clean sheets; drapes on the windows; even a swimming pool out back. The best part, however, was the meals. Because we were in cattle country, every morning featured steak and eggs and rice and beans for breakfast. If we were there for lunch, we'd get steak and eggs and rice and beans again. Then, for dinner, there was a BIG steak and LOTS of rice and beans. We also had rich, dark Costa Rican beer to wash everything down. We soon considered these creature comforts critical to the effort as we began to undertake the more strenuous aspects of the project.

Our mission was to study the disease in horses and in people. Each presented a set of challenges. To bleed horses, we had to bribe Costa Rican cowboys, unless, of course, one was a lot more skillful at catching wild horses than the average Public Health Service doc. The basic rule was that we had to bleed all the horses in a given area because we couldn't just take somebody's word about where the horses with the disease were. That meant bottles of scotch to any-body who could round up a given number of animals, particularly the wild ones in the distant pastures. Often, our problem was not finding someone willing to help as much as it was explaining what we needed him to do, which was no small task. This was real cowboy country: the people we were dealing with carried a rifle and machete on their horse at all times and sometimes a pistol as a personal side arm. We tended to listen very carefully when they spoke. The prob-lem was that until we got used to the rural, country-boy Spanish spoken around Guanacaste—and until the locals got used to our pidgin Spanish—we found ourselves suffering from "a failure to communicate."

One time, Gerry was trying to explain to a rancher that we wanted to bleed his horses.

The guy asked, "Is it going to hurt the horses?"

"No," Gerry assured him, "it won't hurt the horses, but we want all of the horses, not just the ones you're riding. We need unbroken horses, too."

The guy said, "My cowboys have been working all day. We start at five o'clock in the morning and work until two o'clock in the after-noon. It's two now and they're all going home. They're tired."

Gerry said, "Good, I'm glad you can get the horses for us."

Confused, the guy tried again. "No. I can't. The cowboys are all tired and even if I ask them they won't do it."

Gerry closed the sale: "As a gesture of our appreciation, I'd like to give you this bottle of Johnnie Walker Red Label."

Our new friend smiled. "Thank you very much. We'll drink it tonight. Come back tomorrow and we'll help you get the blood."

He went home to catch up on sleep and I tried to keep Gerry from unpacking gear for samples that would not be available until the next day. That's the way it works in the field. You have to improvise. It's not an exact science—nothing I learned at Hopkins and not even something David would have learned at Harvard.

After that false start, though, the guy and his cowboys were very helpful the next day, rounding up the horses and putting them through a chute so we could bleed them. We'd stand on one side of the chute while the cowboy would stand on the other and catch them with what's called a twitch—a loop of rope on a stick they use to catch the horse's mouth and hold it so it can't bite you (although the cowboy was still in dangerously close range). We'd just stick the needle in their jugular, which in a horse is a huge vein you can hit from across the room. The horses were so intent on trying to bite the cowboys that they wouldn't even notice what we were doing. Even hitting the jugular, our needle was a lot less painful than a bite from one of those gargantuan Costa Rican stable flies.

It worked a little differently on the smaller farms. At one tiny farm, after an easy morning bleeding horses on a big ranch, we asked to bleed all the animals and people there.

The farmer asked, "Are you going to kill them?" After we assured him we wouldn't, he let us bleed his wife first, presumably as a test case. It was odd to see it at first, but later I'd see this a lot, particularly in African villages: people and animals are bled in order of importance. In a village where they think the bleeding will be good for them, they'll let you bleed the men, the boys, the adult women, and then the girls. In places where they're afraid something bad will happen, you'll bleed the girls, the adult women, the boys, and then—if everything has gone well up to that point—maybe you'll get the adult men. At this farm, we bled the man's wife and kids and then he let us bleed him. After a little more discussion, he let us bleed his cows and a calf until the only animal left to bleed was their dog, Tarzan. We asked, "Can we bleed Tarzan now?"

Even after all the empirical data before him, the worried man asked, "Are you sure you're not going kill him?"

We looked over at the short-haired mongrel with a lopsided, curved tail, eyeing us suspiciously. "No, don't worry. We won't kill Tarzan."

He dragged Tarzan over by the scruff of the neck and Gerry asked who would help him.

Bill Reeves brashly volunteered, "I don't know how to bleed dogs; show me how."

Gerry said, "It's really easy. You hold him the way I tell you and I'll bleed him." Bill put Tarzan in front of him on the ground outside the dirt-floored, one-room ranch house, put one hand on his muzzle so he couldn't bite, and grabbed a front leg with the other, holding it tight like a tourniquet and extending it so Gerry could see the vein—essentially the same vein used to draw blood from a human. Gerry cleaned the furry leg with alcohol, stuck the needle in Tarzan, and the operation looked like a complete success as blood started to flow.

Unfortunately, Tarzan, like many others in the field that day, had diarrhea. He chose that moment to let loose all over the laces of Bill's boot, which startled Bill. He let go of Tarzan, the needle popped out of the vein, blood squirted everywhere, Tarzan tried to bite Gerry, Gerry fell on his butt, and Tarzan took to the hills. This scene flashes in my mind periodically when I confront a new epidemic. This is what I mean by controlling the externals. You have to be as prepared as possible for every trip out in the field, but you can't be a real rigid personality type or you're not going to make it. In this case, we'd already given the man a bottle of scotch, so we gave him some money and suggested he might want to take his family into town for some soft drinks or beer. Then we went back to the hotel and set the next day's agenda by poolside . . . while Bill cleaned out his boots.

The more demanding phase of the project was our work with human subjects. We'd heard about a town called Cuajinaquil, where there'd been outbreaks of VEE, and thought that would be a good place to study the disease in people.

Up to this point, all our work was done in places easily accessible by the Pan-American Highway. Cuajinaquil, on the other hand, was about ten miles down the escarpment. Travel was via a dirt road, going from highlands to lowlands, and we were getting into the Costa Rican rainy season. We had three vehicles: two from MARU (including the Black Mother) and one from the Costa Rican Ministry of Health. Our vehicles were not in the best of shape even before we started downhill in mud. As far as maintenance and replacement goes, the PHS bureaucracy sets the number of miles a vehicle is supposed to be driven before it's turned in. Of course, our mother institution, the NIH, calculated the miles across paved roads in a Maryland suburb. We counted our miles across creeks, ditches, sand, and mountain gorges. But you work with what you have. When we started down to Cuajinaquil that first day it was already drizzling and the locals warned us we might not be able to make it back out. It got worse and worse the closer we got to town but we were focused on our mission. Cuajinaquil was a town of about 800 people who all lived off one main dirt road. At the big corner in the center of town there were two general stores, diagonally across from each other, a telephone booth—which was the one phone in the town—and a tack store or a blacksmith. We checked in one of the general stores and confirmed the rumors we'd heard in Guanacaste. There were cases consistent with human VEE in town. We suspected these were some of the first naturally occurring human cases in this epidemic and we wanted to prove it. This also made the town an ideal place to study the disease. With recent onset, we had acute patients to collect samples from. We also had people who'd probably been previously exposed and had sero-converted, meaning they had antibodies in their blood and were now immune to the disease; and we'd be able to sample the whole town, calculating the ratio of cases (people who got the disease) against people who sero-converted. Most viruses produce a wide range of effects and many more people may be infected than actually come down with overt disease. Therefore, we speak of case-to-infection ratio.

Our excitement at finding this research gold mine was tempered, though, by the weather, which turned into a Central American rainy season storm, dumping buckets of water on us. We knew we had to get out of there, but by this time the road had gone from bad to worse; the mud was thick and whole sections were washed out. On our way out, the Black Mother got stuck and we tore up the transmission trying to get it unstuck. Our other International soon joined it, and the Toyota from the Costa Rican Ministry of Health was similarly overwhelmed—we used some of our needles to fashion a new cotter pin to hold the carburetor together so at least we could get that one moving again. The rest of us sat there, soaking and stewing in the middle of the road, until salvation came in the form of a large Mercedes four-wheel-drive cheese truck that was somehow able to cut through the mud. The driver felt sorry enough for us that he agreed to haul us out of that mess.

Back in Guanacaste we faced the locals, who looked very concerned for us and were kind enough not to say they told us so, depressed, embarrassed, and more than a little chagrined at the fact that in one afternoon we'd managed to annihilate our entire fleet. Luckily, there was a good mechanic in town who had a tractor, which is the one vehicle that can go virtually anywhere. He and the cheese truck driver took us back down the hill, pulled out our vehicles, and the mechanic replaced the broken transmission on the spot.

The truck driver and I got to talking and he told me the U.S. Peace Corps was a wonderful idea, but "it's not fair."

"Why not?" I asked, practicing my Spanish.

"We have four of them in town right now," he explained. "The male volunteers sleep with our women and the female volunteers do not sleep with Costa Rican men."

From there he introduced me to a difference in perception of Latin men, at least in that town, which was that they like their women somewhat bulkier than you might imagine. He described how he admired the ample proportions of one wonderful Peace Corps volunteer as she rode her bike through town. I wondered if she had any idea how the locals felt about her.

From that day on, we worked from a staging area we could reach with the vehicles, hiking down to bleed the people and then back up with the samples to the staging area. In 90-degree weather and 99 percent humidity, we didn't fare that much better than the trucks did. All you had to do was move to break a sweat, and although we were all pretty young and reasonably fit, the walk out was terrible. Two steps forward, one step back, uphill in mud, trying to make sure the test tubes of blood didn't break.

I also learned something about reading local political dynamics. Pretty early on I noticed that whole sections of the town would not agree to be bled. After asking around, I finally found one guy who would chat with me and he explained what was going on. It seemed that the two families who owned the general stores in town were in a long-running feud and the whole town had divided into two camps. Simply by visiting the one general store that first day we had alienated a whole section of town. So the next day we shopped at the other store for lunch, which was apparently enough to settle the score. We'd made it clear we were not allied with the one family, so the other family and their friends canceled their vendetta against us. With the exception of the widow Lopez's daughters—three beauties the old woman was bound to protect from all men but especially stinking, sweating foreigners—we were eventually able to bleed nearly everyone in town.

After taking our blood samples, we would stop at one of the general stores for lunch. The selection there consisted of canned tuna fish, canned pineapple, and some sodas. By the second day, the four of us had completely depleted the supplies of the first general store, which gives an idea of how close to the bone many rural Costa Ricans live, for Cuajinaquil was not an exceptionally poor town. We'd continue working until close to two, by which time we had to start our hike back up the hill before the severe rains came. At two o'clock every day we knew it was going to rain like a cow pissing on a flat rock. We'd usually reach the truck exhausted—often taking turns dragging the most tired among us the last few steps—minutes before the rains hit. Even when we made it to the trucks in time we often had to stop on the highway until especially bad rains would let up a bit. Then we'd go back to the hotel, shower, and collapse, trying to get an hour's sleep until five or six, when we'd get up and process the day's samples, have some beer and steaks and rice and beans for dinner, and go to bed so we could get up and do it all again at five o'clock the next morning.

My experiences kept driving home to me that field research is just not something you can learn in a classroom or a laboratory. While we were out there, CDC sent out an EIS officer who was very well educated but had no real experience in the field. He listened and observed what we were doing, dragging our asses up that hill in the mud every day, before finally declaring our strategy too labor-intensive. He suggested we use the phone book to conduct a random telephonic survey of all the houses in the town, asking people about any cases in their household. We didn't bother pointing out that the entire town had only one phone.

If you can appreciate the rewards of the work, though, there's nothing like it. We all bitched and moaned, but nobody seriously complained, because we knew this was groundbreaking work. We were getting blood samples we could convert to new, unique clinical data that would really help people.

In the middle of the week, Gerry and some of the other guys broke off from the group and began identifying and inoculating horses. In a combination of fast footwork and an enormous stroke of luck, they were able to hit some ranches just before the epidemic swept into the area. Horses given the vaccine only eight days ago were protected. As a matter of fact, the field crew had been taking samples and could show that the last new infection occurred only three days before vaccination. It also showed that the vaccine was safe in the field—even in pregnant mares. That essentially set the stage for the vaccine effort later carried out in the rest of Latin America. While we were out there, sometimes I could focus on the big picture—that we might be gathering information that could eradicate disease—and sometimes I had just enough energy to taste how good the beer was and appreciate a finally full stomach.

It was only after our studies were long complete that our own Department of Agriculture began to think about vaccinating horses in the United States. Before then, the USDA cringed at the idea of using a live-attenuated vaccine in the field, even if it had been tested extensively in the lab. For whatever reason, there was no vaccination campaign in Mexico as the disease pushed north at a rate of 2–3 miles per day. But the USDA didn't vaccinate against VEE until the disease was already in Texas, threatening the big ranches. As a result, there were hundreds of deaths in horses and hundreds of cases of disease in humans—all of them preventable. Sometimes it's easier to get the chief of a rural village to understand the value of battling these viruses early than it is to convince the more "sophisticated" leaders in our own government's agencies.

And, in perhaps the biggest irony of all, two decades later, when we had the technology to conduct genetic testing on the VEE virus strain from the outbreak, we found that the source for the epidemic was almost certainly an improperly inactivated vaccine in use in Central America at the time.

When I got back to the lab and had time to think about what we did, I realized that the greatest threat to American citizens from VEE—as well as the virus—came not from the microbe itself, but from the giant bureaucracy afraid of committing to an action. We lucked out that time because the Public Health Service had a lab in Panama with people who had experience with the disease and were ready to work with the governments in other countries affected and deal with it on the spot.

Today, now that a lack of funding has closed MARU and other labs like it, I wonder who will be in these places to spot the threats of the future. Who has the access and the experience to deal with these diseases before they reach our country? VEE proved to me that we can deal with these diseases now, where they come from, or we can wait until we see them in our backyard and start from square one after our own citizens start getting sick and dying. And that's a crying shame.

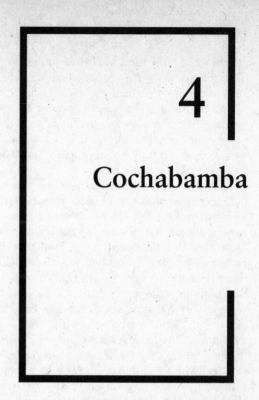

4

Cochabamba

WHEN I RETURNED to MARU from rural Costa Rica, it was time for me to get back into my laboratory project. I'd seen firsthand how field epidemiology fit into the study of viral diseases and was ready to learn how the laboratory brought the picture into sharper focus.

I was sitting in my office one day, kind of staring at the wall and looking at my lab book and wondering what to do next, when Karl Johnson came in and said he had been at NIH in Bethesda, Maryland, and visited a guy named Bob Purcell. And Purcell had talked to him about a new finding in the hepatitis field called Australia antigen. It seemed that Baruch Blumberg, who later received the Nobel Prize for his part in the discovery of Australia antigen, was taking blood serum from hemophiliacs who had been multiply transfused—that is, who'd received serum from a number of different donors—and using it to test for antigenic differences in different populations.

In other words, if you inject another species, say a rabbit, with your serum, he's going to make antibodies against everything, so it's going to be a very complex analysis to figure out exactly what's reacting against what. But if we take your serum and inject it into me, I'm only going to produce antibodies against the many fewer things that are different between the two of us humans. These differences in sera within the same species usually turn out to identify genetic differences between the two.

Blumberg identified a particular antigen that was found only in certain people's blood. He first found the antigen in the blood of Australian Aborigines, hence the name. He didn't know what it represented and initially never dreamed it had anything to do with disease; all he knew at that point was that here was something in some of these Aborigines' blood that wasn't in other people's. Then he found it in increased frequency in kids with Down's syndrome, and then in leukemia patients. As it happened, his lab technician came down with hepatitis and the antigen appeared in her serum. So he started testing other hepatitis patients and found in them the highest prevalence of all.

In the late 1960s we talked about two kinds of hepatitis: infectious hepatitis, the kind I had, with a short incubation period and a milder disease (although it didn't seem mild to me at the time), which was not known as something you carry with you your whole life. That's now called hepatitis A. The other kind, then called serum hepatitis and now hepatitis B, was much more serious, sometimes life-threatening, and had a longer incubation period. One contracted it either from a transfusion or from an improperly sterilized syringe. Although the virus had only just been discovered in the late 1960s, many people today know of it as the condition that forced popular country music singer Naomi Judd to retire. The link Blumberg had made, and on which Bob Purcell at NIH was following up, was that Australia antigen was a marker for hepatitis B.

The hepatitis B virus infects only humans, and its symptoms can range from the virtually unnoticeable to the completely debilitating, with severe liver damage. Most hepatitis patients develop an effective immune response and successfully get rid of the virus, but some become lifelong carriers, capable of infecting others—for example, if their blood is given to someone in a transfusion.

Karl set two vials of reagents on my desk and said, "I think this is real; see what you can find out about Australia antigen in Panama." I began to study acute hepatitis patients in Panama and the Canal Zone but soon became more interested in trying to understand the mystery of Australia antigen in the blood of long-term carriers. Blumberg had determined that in a healthy Caucasian population, for example, less than 1 percent of blood samples tested will have the antigen. But there were a few populations, like the Aborigines, where it turned up with much greater prevalence. The studies thus far had indicated a high proportion of carriers in certain tropical popula-tions. And living and working in such a tropical area, I realized that my studies required field research as well as lab work. It would be up to me to explore rural locales in Central America, testing the "na-tives," my new neighbors, for hepatitis B antigen.

Not only is hepatitis a serious and often deadly disease throughout the world, but a knowledge of the subtleties of genetic differences between infected hosts is a critical piece of the puzzle in understand-ing why a seemingly identical virus causes serious disease or death in one individual or group and little reaction or none at all in others. Is the difference in the virus itself, in the host, or some other factor we haven't identified? As we saw both in Ebola and hantavirus, getting a handle on these sorts of variables is a big part of what epidemiology and public health are all about.

There were three Indian groups in South America selected for the study. I began with the Cuna, who happened to live in one of the most enchantingly beautiful places on earth: the San Blas Islands, a small archipelago off the Atlantic coast of Panama. The Cuna offered a fascinating culture, not to mention some of the most picturesque, smooth white-sand beaches, lush coconut trees, and the most fantas-tic coral reefs for diving imaginable. I gathered the scientific equip-ment I'd need, making sure to pack my swimsuit and snorkeling gear, too.

The friendliness I encountered upon my arrival, surrounded by colorfully garbed Cuna of all ages, watching with beguiling smiles of curiosity, gave no hint of the difficulty I'd face trying to work with them. Friendly is one thing, receptive to being poked with a needle to give up one's blood to an odd-looking stranger is another. I went back to my college professor Trenton Wann's lesson: if I wanted to succeed in my mission, I had to work within the framework of their culture. I thought of my experience in Cuajinaquil, side-stepping community politics by simply purchasing lunch at another store. Here, I realized, I was dealing with a tribe who did not even recog-nize the authority of "their" government in Panama, much less un-derstand what giving up a little of their blood would accomplish for them or for science; what was science anyway? I had to learn their rules.

As I knew the Cuna were a group strongly structured around their chief, I made it my business to pay homage to—and plead my case before—the tribe leader. I'd learned something from our efforts in Costa Rica and made arrangements for an interpreter to accompany me (my Cuna was nonexistent and my Spanish still poor, and while I could bet the Cuna were more proficient in Spanish than I, it was sure to be of a rural variety not necessarily understandable to me). My interpreter provided an interesting diversion. An albino, he was immediately recognized by the Cuna as a "moon child," though I'm not sure whether this helped the project any.

To see the chief, we entered the largest of many thatched huts with sand floors and were soon joined by several Cuna who carried in chairs for our meeting. The chief was an elderly man, and I was surprised to find him dressed in Western garb, including suit, tie, and hat.

Our meeting consisted of an interesting exchange. I would take thirty seconds to explain myself in Spanish to the interpreter, who then took about sixty seconds to relate it to the chief in Cuna. The chief, in turn, spoke Cuna back to the interpreter for sixty seconds before the interpreter turned to me and said, "He says no." It took a little while to uncover the source of the resistance, but we finally realized the chief was holding out for a reason: he wanted measles vaccine for his people. Once I knew this, I coordinated with the Panamanian government physician, Felix—an Argentine doc who'd found the San Blas Islands so agreeable he opted to stay and become personal physician for 35,000 Cuna Indians. MARU bought the vaccine, Felix brought the equipment, and we bled as we vaccinated. We also held a small clinic.

Whenever we do studies that involve collecting blood from people, we must have them sign a consent form indicating that they understood what was to be done and they freely and willingly gave their consent. This is a sticky issue. I never knew how much of my explanation made sense in translation, both linguistic and cultural. Certain issues were easy to deal with: I was able to assure the Cuna I was not taking their blood back to Panama to be sold. But when I said I would test it to try to understand more about the health of their community so that ultimately the government of Panama and their own leaders would be able to plan to improve their health, I was less sure they understood. I never offered specific or direct benefits to them individually ("This will make you well . . .") but stressed the long-term general benefit to their people. Even if it weren't mandated by my employer and the government, I would have tried to explain things. I believe very strongly we owe people the truth about what we do—with no unrealistic promises—so they can make their own choice of whether or not to participate. Parenthetically, I also believe that peoples all over the world owe it to themselves and others to participate in reasonable studies to find out more about diseases and to take part in control activities that can protect their communities.

When I got their blood back to the lab and tested it, I was surprised to find no evidence of chronic hepatitis B infection among the Cuna. Bill Reeves and I later looked for other viruses (I had so looked forward to snorkeling there), but found a lower incidence of infection overall in these Indians. We explained these results in terms of the Cuna's excellent hygiene. Since the islands had no water, they caught drinking water when it rained or sometimes went to nearby streams on the mainland located in areas so sparsely populated they were thoroughly nonpolluted. And, of course, for wash water and waste they had the biggest flush toilet in the world! One morning while birdwatching I picked up an odd image through my binoculars—a group of Cuna Indians performing their ablutions, backs to me, in the morning waves.

As hard as it was to admit the need to move on, I learned another real-world truth from the experience. In science, you have to be able to abandon a preformed hypothesis—no matter how intellectually attractive it may be—in the face of contradicting data or indicators. Sometimes you learn just as much, or more, disproving a theory as you do proving it. That's one of the benefits of the scientific method. Unlike as in other fields of endeavor, we usually don't have "failures." We have experiments which prove or disprove hypotheses.

After a stint studying other Indian groups, such as the Choco and Guaymi, who reacted differently to hepatitis infection, we were able to show that among the Guaymi the carriers resulted from genetic differences in the infected humans. My research was interrupted, however, by a dramatic development that was to change the course of my career: an outbreak of hemorrhagic fever in Bolivia in February 1971.

It's a tribute to Karl Johnson's diplomatic abilities that he was called by Bolivian authorities as soon as they recognized the beginning of an epidemic. In those days, Bolivia was having revolutions at a frequency of greater than one per year, so you'd go down and negotiate how you were going to work with Bolivian health authorities and by the time you got back there'd been another revolution and your contacts were out the proverbial window. There simply wasn't enough scotch to keep bringing gifts to the regime-of-the-month. As a result, MARU didn't have any active programs with the country at the time. But Karl had been able to regularize contact to the extent that whoever was in charge there knew he was the man to call with any serious public health concerns.

It turned out that people in Cochabamba were getting very sick, and some were dying, from what looked like Bolivian hemorrhagic fever (BHF). This was unusual. BHF, caused by a virus known as Machupo, had never been known to occur in that region of the country. Cochabamba was located in the highlands, and all previous cases of BHF had occurred in the savannas in the interior of the country, known as the Beni province.

Although it was an anomaly for the disease to strike outside the Beni, previous outbreaks had taught health authorities to take hemorrhagic fever seriously. A 1960s outbreak in the town of Orobayaya caused more than 100 cases and left 44 people dead in a town of only 600 inhabitants. But the worst epidemic hit in 1963–64 in San Joaquín. It ravaged the city for two years, infecting 637 people from a population of 2,500—including a team of American scientists who had to be evacuated from the town. Altogether, 113 people died there.

There was no cure for BHF—nor is there yet—but there were two positive results from that outbreak that would hopefully help us eradicate the disease one day. First, scientists, led by Karl Johnson, successfully isolated the virus from the spleen of a young boy, Juan Carvallo. Second, they were finally able to link the virus to a specific reservoir: *Calomys callosus,* a rodent about the size of a deer mouse that served as reservoir for the Sin Nombre hantavirus in the southwestern United States in 1993.

When I describe Karl Johnson as something of a virology guru, I have contributions such as his work with BHF to back up that statement. At the time of the San Joaquín outbreak, there was no knowledge whatsoever of the family of arenaviruses. All that scientists knew of was one lone, orphaned virus, lymphocytic choriomeningitis virus (LCM), that didn't share characteristics with any other known viruses. In fact, although there were many classical bacterial and rickettsial diseases such as plague, relapsing fever, or typhus that used rodents as hosts, there were no other known rodent-borne viruses that caused human disease except for an obscure, unclassified virus recently identified in Argentina that caused hemorrhagic fever.

Karl first hypothesized that the virus responsible for the San Joa-

quín outbreak was related to the rodent population. He devised a
plan to test his idea that was both insightful and bold, and I'm sure a
lot of people—both residents of San Joaquín and some scientists—
criticized it. They couldn't imagine the disease came from a virus
carried by a common local rodent. Karl split the town right down the
middle and on one side had all the rodents trapped out of the houses.
They were able to count almost 3,000 rodents trapped, with some
unknown number poisoned and not recovered—thus there were even
more rodents than people. The first half of the town was started on
May 1, 1964, and the second half on June 15. The decrease in cases
in the first half began after May 15, and after June 28 there were no
new cases in the whole town!

From a scientific standpoint, it's critical to have a control group in
an experiment like this—especially when you're working with a disease that can seem erratic in its behavior: you can't tell which household it will strike next. From a PR standpoint, this tactic put Karl in a
potentially no-win situation. If it didn't work, he disrupted a lot of
households and diverted a lot of time and resources catching and
killing rodents; if it did work, he left half the town exposed to a
disease with up to a 30 percent mortality rate. But a big part of
epidemiology and science in general is having the courage of your
convictions, because it doesn't matter what your theories sound like
if you don't have hard data (in this type of study, from a control
group) to support them.

Everyone saw firsthand that in the half of town where they started
the trapping, there were no new cases after the incubation period
passed. Then the same thing happened on the other side of town after
the rodents were trapped, offering dramatic proof that *Calomys callosus* was the rodent reservoir of the virus. It also showed unequivo-
cally that you can control urban BHF by removing rodents. It convinced residents and the government in a way they couldn't deny and
would not forget. And when the epidemic was over, Karl headed
studies to rule out other species as factors: they tested almost 30,000
arthropods looking for virus, for example, and found none, in addition to testing other rodent, bat, and insectivore species. Compare
this with the original survey, which found Machupo virus in fifteen
out of thirty-one *Calomys* tested. Now we know that in an epidemic
situation it's common to find up to 30 percent of the rodents infected
with the virus.

Calomys is a field rodent, and just as in the old fable, field mice
and town mice are different. Historically, people who caught BHF

worked in the cleared forest areas in the Beni and came into contact with the rodents there, causing the sporadic cases that were occasionally reported. The outbreaks in Orobayaya and San Joaquín in the 1960s, while remarkable in terms of the number of cases and deaths, were consistent with the sporadic cases because both cities were located in the endemic area—the lowland savanna the Bolivians call the Beni. For some reason—a population explosion in rodents, increased clearing of land, decreased food source in the rural areas for the rodents, lack of predators and rodent competition in the towns—the rural *Calomys* moved into the towns and people started getting infected.

But in 1971 there was no such logical explanation for what was happening in Cochabamba, well outside infected *Calomys* territory. In the lowlands, after Karl's trapping experiment, towns had been vigilant about rodent control. The scattered cases that arose since came from people who traveled into the fields, tending a herd of cattle or a yucca garden outside of town. When rodents invaded the towns or if people in the towns got sick, people knew what to do: get rid of the rodents and you'll prevent the disease. In the endemic area, the disease was basically under control.

Cochabamba was 350 miles away as the crow flies, and you'd have had to be a crow or a Cessna to get there that directly from San Joaquín in the Beni, because there were no roads. Cochabamba was Bolivia's third-largest city, high on the slopes of the Andes. The highlanders lived on the altiplano surrounded by Quechua and Aymara Indians and called themselves Camba. The highlands contained the miserable mineral mines that killed many of the Indian men who worked them by age forty, and the now exhausted silver mines that had made the mining town, Potosí, a Spanish-language synonym for vast riches. The Beni region and neighboring Santa Cruz were rich, expanding farming and ranching areas with oil beneath the surface and some substantial cocaine trade as well; the people there were the Kolla. But the Beni also had Machupo virus infecting many of its rodents, whereas the highlands had been spared this particular plague.

From preliminary reports we heard, the disease presented symptoms of BHF with a twist: people were jaundiced, and it appeared the mortality rate was higher than expected. In previous epidemics, there was a tendency for the disease to cluster in households but it wasn't spreading from one person to another. It was what we refer to as an "infected house" phenomenon. Several members of the same family might go to the hospital sick because of infected rodents in the house, but the doctors and nurses who treated them were relatively safe because the patients were infected by the rodents, not by each other.

When Karl got the call in 1971, though, the disease sounded more frightening. The rapid person-to-person spread suggested something was different, perhaps even an aerosol spread like influenza. Imagine, for instance, if AIDS were transmitted through the air rather than requiring intimate contact. The entire human population of the globe could be infected by now. In the laboratory, virus-containing aerosols are easily generated, and Machupo (with its relatives) is a notorious and lethal hazard. But we had never had an indication that patients produced dangerous amounts of aerosols.

In Cochabamba, one patient who entered the hospital severely ill with viral hemorrhagic fever somehow infected two of her relatives and two nurses before she died. Whether it was BHF or not—caused by Machupo or another virus—this was a real epidemic. People were getting sick very quickly and there was a high mortality rate: the nurses grew ill about a week to ten days after the index case came in and three of the four secondary cases died.

The stakes are always higher when a mystery disease spreads in a hospital. When a community realizes the hospital is making people sick, the fear factor goes through the roof as morale among health-care professionals plummets. People in Cochabamba knew that the folks in the lowlands got this disease and that rodents somehow gave it to them, but they thought the Cambas in the highlands were safe. Now it seemed that nowhere was safe.

The first question Karl and I discussed was whether this was BHF or not. The arguments against it were that it took place outside the Beni and seemed to have person-to-person transmission on a scale never seen before with BHF. There's a big difference between a recovering scientist passing the virus on to his doting wife during convalescence and health-care workers being exposed through strangers in a hospital setting. And the symptomatology was somewhat different. These patients in Cochabamba were developing jaundice as a consequence of liver damage, something almost never seen with BHF. But the difference in altitude could be another variable. In short, we couldn't tell what was going on there.

We did reason, though, that if it was BHF, the obvious thing for Karl to do would be to send someone known to be immune through previous exposure and recovery. Well, Karl was head of the lab and up to his armpits in VEE, so to speak, so he couldn't go. Patricia had developed another illness, so she couldn't go. And the other immunes in the lab were good technicians but weren't physicians. Karl reasoned that someone had to go, we really didn't have clear indication this was BHF, and anyway, we knew from the painful example of Einar the limits of hedging your bet with immune researchers. Thus, although it was not lost on anyone (least of all me) that I had not had Bolivian hemorrhagic fever, once again my dance card came up. When Karl asked me if I wanted to go, I said, "Of course."

My youthful sense of invincibility wavered a little after the excitement wore off and I started packing my things. But a well-known syndrome hits you as a young physician and epidemiologist faced with an unknown, possibly dangerous situation. You become a fire dog: the bell goes off and you jump to attention, ready to go. Over time, with experience and seasoning, you get fire-dog fatigue, not to mention an appreciation of your own mortality. You reserve the all-out emotional response for the biggies. You don't get pumped for every case that comes along.

But this mysterious epidemic in Cochabamba hit everyone the same way, fresh young pup to experienced fireman. At that point no one in the world knew terribly much about these viral hemorrhagic fevers. Here, we had a chance to observe the progress of one in real time, the first step toward controlling a disease. In this case that opportunity could lead to either greater understanding of a disease we already knew about but hadn't seen in these conditions before or the start of data collection on a brand-new disease. In either situation, we would have new and exciting information and eventually it might even lead to lives saved later down the line. These thoughts—coupled with the general exhilaration of going to a completely new area—kept my enthusiasm level up and any personal concerns well at bay.

Unfortunately, these positive thoughts were not so prevalent around the lab and at Gorgas Hospital, where people were a lot more concerned about my safety and welfare, frankly, than I was—one person in particular.

One day a year before, while searching for abnormal liver function test results from hepatitis patients at Gorgas, I'd run into a tall, slender, attractive woman working as a technician there, Norma Quintero. Two years older than I, she was a graduate of the University of Panama, and proficient enough in English to allow truly interesting conversations. We started out as friends. She had two children by a previous relationship with a prominent Central American leftist

playwright and was in no hurry to get recommitted; and I was still
smarting from my own divorce, although I'd finally started social-
izing again.

After a while, though, we started dating on a more serious level.
Her children and I got along well. Her daughter, Antígona, was five,
and her son, José, known as CheChe, was four. We spent quiet
evenings at her place, went out together with the kids or dropped
them at her mother's house for an evening out alone together. Before
I knew it I was spending Sundays at her mom's house, hanging out
with her three brothers and a sister and their families.

Compared with other Panamanian women I'd dated, Norma
seemed much more Americanized and her family more comfortable
with an American in their midst. She'd been quite a campus radical
when she was younger. Working in the Canal Zone, becoming seriously involved with a foreigner, these were just the most recent developments in a life that was already breaking a mold. In Panamanian
terms, she had always been something of a rebel. She was smart and
adventuresome enough to actually look forward to moving to the
States with me when I returned in a few years' time. On June 22,
1970, Norma and I got married in Panama.

Now, just eight months later, the honeymoon was over. I went to
the hospital to let Norma know I was going to Cochabamba, but her
boss in the lab had beat me to it. "Gee, Norma, we're really sorry to
hear Pete's going to Bolivia. We really liked him. We'll miss
him . . ." She was not happy.

Karl and I discussed what kinds of precautions I should take. This
disease sounded as bad to us as anything we'd heard of. Karl decided
I should take respirators with me. I was to use them at all times if I
thought there was *any* evidence of aerosol transmission. The assumption was if I got sick I would come back to the Canal Zone for
treatment, although in view of the speed and severity of the disease's
course, the trip could kill me. And if I did make it back the best they
could do was reactive treatment, responding to individual symptoms,
since there was no treatment for either BHF or this potentially new
BHF-like illness. But at least I'd be somewhere where my new wife
could hold my hand in isolation—if they let her in.

Fortunately, I was young and immortal in those days, and getting
equipment ready for the trip, including liquid-nitrogen freezer, autopsy instruments, and blood-drawing gear, made me busy enough to
keep me from worrying as much, perhaps, as I should have. I caught
a flight to La Paz on a four-engine prop in the famous fleet of Lloyd's

Aero Boliviano. But as soon as my plane arrived I confronted my first challenge. At thirty-one, I was in reasonable shape—fit enough to hike through mud slides in rural Costa Rica and snorkel around the San Blas Islands. But La Paz sits at an altitude of about 11,000 feet and I was used to the air at MARU, virtually at sea level. I got off the airplane, walked down the stairs and across the tarmac, and had to sit down to catch my breath as soon as I made it to the terminal.

There was a small entourage waiting to meet me there, luckily by a bench. The Bolivian physician acting as official representative for the Ministry of Health introduced himself and an associate from Cochabamba who would help with the work there, Rudolfo Mercado. As I greeted Rudolfo, I recognized a third man, a gringo like me, Bob LeBow—a former medical school classmate. I hadn't seen him since med school and had no idea he was working as a physician for Peace Corps volunteers, much less living in Cochabamba, Bolivia.

Bob was tall, skinny, with a perennial heavy five o'clock shadow and absolutely hilarious. A reliable, engaging guy, he was ideal to work with under the circumstances. And his wife was a ham radio operator, another plus. This would prove critical later, since communications with La Paz were somewhat sporadic and I needed some way to get in touch with Karl. Although Cochabamba was the third- or fourth-largest city in Bolivia, the phones were strictly local. You could call throughout the city, and sometimes to La Paz, but never out to Panama.

It always takes one good clinician to spot the pattern of illness in an epidemic. In this case, that clinician was my old friend Bob LeBow. In Cochabamba, it was his role to take care of the volunteers in the area and learn about local diseases. In his spare time he made rounds at different hospitals in town. This taught him about local diseases so he could better treat the volunteers and it allowed local physicians to learn from his experience.

LeBow described the history of the outbreak for me. It started with a twenty-year-old single nursing student who lived in Cochabamba, but recently returned to her hometown in the Beni, a village named Fortaleza. There were no signs of an epidemic there, but by the time she left for the city of Trinidad on her way back to Cochabamba, she started feeling ill. She arrived in Cochabamba on January 15, having been sick for three days, and checked into Seton Hospital. After developing central nervous system symptoms and hemorrhage consistent with Argentine or Bolivian hemorrhagic fever, she died about two weeks later, on January 28.

In the last two days of the index case's life, her father and aunt both assisted in her care. Like many Latin Americans, they were both very expressive in their love and grief over the young girl's fatal illness, weeping and hugging her, when presumably she had the highest levels of virus in her blood, unwittingly exposing themselves.

Just one week after the girl's death, her thirty-three-year-old aunt became sick and was later admitted to another local hospital, Viedma Hospital. She died about two weeks later, on February 23. The girl's father, in the meantime, had returned home to the Beni, where he also got sick. He checked into the hospital in Trinidad (in Beni province, not far from Fortaleza) and died there on February 19. So that was the epidemiology we were working with.

Seton Hospital was one of the places Bob LeBow visited regularly, as was Viedma. He observed the index case in Seton and later recognized her aunt at Viedma, exhibiting signs of the same dramatic illness two weeks after the first case. He knew the girl's father had been in similar contact with his dying daughter and tracked him down, contacting health-care workers in Trinidad. Unfortunately, he was not able to see this patient, but did get confirmation from doctors there that they believed he was suffering from a severe hemorrhagic fever which caused his death after approximately a two-week period of illness. This was enough to give Bob the beginnings of the "pucker factor."

The next two cases were frightening because they were nurses, one of whom had absolute minimum contact with the patient. The first, twenty-eight years old, had handled and touched the index case, supervising her care throughout her hospitalization. Even with this patient contact, though, she had done nothing approximating classic exposure to viral hemorrhagic fever; she had not pricked herself with a needle, for example. She was admitted as a patient at Seton on February 12 and died just one week later.

The second nurse's illness was more disturbing because she had not even touched the patient. A visiting nursing student from Tarija, another city in the highlands, she was at Seton for training. She'd been present in the index case's room when the first nurse demonstrated how to change the bedsheets but she never touched the sheets. She'd had no contact with the patient at all: didn't carry a needle, didn't give any medication, didn't even carry the sheets out. She was just standing there in the room, which made her illness the kind of thing we only see with aerosol transmission. She complained of gastroenteritis-like symptoms as early as January 25 (at the height of the index case's illness), but said she felt better by February 2, so she left to return to Tarija. Local health authorities there contacted Seton Hospital in Cochabamba when she was hospitalized on February 11. The nurse recovered after suffering an illness described by her doctors as severe.

By the time I arrived, there'd been news of a possible sixth case: a thirty-two-year-old pathologist at Seton who cut his finger performing an autopsy on the nurse who died there. It took him only four days to develop symptoms of high fever and malaise. His illness was particularly devastating to the Seton Hospital community, who'd already lost one of their own. By all accounts, both he and the dead nurse were among the most popular, best-loved workers there.

As Bob and I took stock of the situation, we realized we had a real good news/bad news scenario brewing. The good news was no further cases had been identified except the pathologist and we were far enough into the next incubation period so that we probably should have seen more by now. We were confident that if there were other cases, we surely would have heard about them, since he had been in touch with local health authorities throughout the affected area and everyone was pretty paranoid.

The bad news was that when we counted up possible cases of exposure, it added up to a whole lot of people. A total of nine people lived with the secondary cases who got sick in Cochabamba; the index case and the nurse at Seton Hospital stayed in private rooms but were attended by at least thirty-two nurses and other hospital workers, plus seven doctors; the index case's aunt was cared for in an open ward at Viedma Hospital where she had direct contact with eleven nurses and six doctors; autopsies had been done on two cases with full protective gear of masks, gown, and gloves worn, but two people had been nicked by needles and one person (the sick pathologist) cut in the process. And at that point, at least eight lab technicians were identified as having analyzed blood samples from the various patients.

These were just the people we could count. If the virus was aerosol-infectious, as seemed to be indicated by the student nurse's illness, uncounted others were possibly exposed. If there were going to be more cases, there could be a lot of them.

In terms of the bigger picture, you'd have to count as more bad news the fact that we couldn't really figure out exactly how, where, or why the epidemic started. The index case and one of the secondary cases—her father—spent some time in the Beni, but there was no evidence of illness in the towns they stayed in. Nor did it appear that they spent any time in the fields where infected rodents were prevalent. The two nurses lived, worked, and traveled strictly in the highlands. And of course there was the disturbing issue of possible aerosol infectivity.

As we sat down to sort things out with our Bolivian associate, Mercado, we knew that the town around us was in a state of panic. Cochabamba had never seen a disease like this before and now it was killing their nurses and doctors! As with any outbreak, there were political elements at play here as well. Even though there is a cultural bias—the people who live in the highlands claiming they don't care about the people who live in the lowlands, and vice versa—when national election time comes up, already shaky governments can stand or fall over things like epidemics. And this outbreak was getting a lot of publicity in La Paz.

Cochabamba itself was a microcosm of the kind of ideological infighting that afflicted the whole of Bolivia. There were three hospitals in town, and where you worked made a statement about your political beliefs. If you worked at Seton Hospital, you were a pro-American, right-wing gringo sympathizer, pro-Catholic Church and against liberation theology. You had to be to work there. Seton had the best equipment (and private rooms) because it received money from contributing Catholics in the United States. It cost the most to stay there, although they did charity work. Viedma was less expensive but didn't have comparable facilities. Those who worked at Viedma were intensely nationalist and often anti-American. And those who worked at the Social Security Hospital were Communists. Patients went to whichever hospital they could afford, but the medical staff chose the hospital they worked at with a political bias.

The political climate only exacerbated an already tense situation. Each group was trying to use the outbreak to their own advantage in an effort to discredit the others. No one wanted any of the sick to die at their hospital. The truth was frightening enough, but the papers were full of rumors fueled by this infighting. One said that there were ten cases of hemorrhagic fever in Viedma; another announced that all the student nurses at Seton (where they had a major nurses' training program) were bleeding from their noses, feverish, and dying. With all the misinformation, the city's residents were in an absolute uproar.

LeBow had been there long enough to appreciate the politics, and Mercado had grown up in it. Now in his forties, Mercado was a very experienced doc who'd ridden donkeys into the most distant corners of Bolivia to investigate plague and yellow fever. Although he'd never dealt with BHF, he was a tough, down-to-earth guy who knew his way around the politics. Their expertise was critical when we began to survey how many people truly had contact with the patients at the different hospitals.

If we were working this epidemic today in the United States, we'd put all the potentially exposed under surveillance and have them come in once a day to have their temperature taken. But back then, we were sort of making up the procedures as we went along. So after we identified everyone we could, we drew blood samples and told them to report back if they got sick. We also designed a questionnaire to determine the extent of each person's exposure, from simply visiting another patient at the open ward where one case was treated to accidentally sticking oneself with a contaminated needle.

Since at this point the only person presumed infected from aerosol transmission was also our only survivor—the nurse in Tarija—I felt it was critical that we interview her extensively. Aerosol transmission would represent such a tremendous departure from all previous incidents with BHF that I felt secondhand reporting was not enough; we needed to hear from her exactly what kind of contact she'd had with the index case.

But she was in Tarija and the only way to get there quickly was to fly out on a DC-3, with scheduled flights leaving only on Tuesdays and Thursdays. We wouldn't have enough time with the patient if we both went and came back on a Tuesday, so the best-case scenario was to go on Tuesday and come back Thursday. But like most other travel in Latin America, this plan couldn't be confirmed even if the flights left on time. The plane had to fly through a mountain pass, landing on a runway in line with the pass. If the winds weren't right or it was too cloudy, the plane couldn't land. They couldn't radio ahead to confirm conditions, and even if they could, the winds were changeable enough so that you could depart only to arrive and find you couldn't land. You pretty much had to get on the plane and cross your fingers. None of the three of us could afford to take that much time at that point, so we briefed a local doc, primed him with questions, and sent him off to interview the surviving nurse and those who'd attended her. When he came back he confirmed there really had been no direct contact, so we had to regard her infection as probably due to aerosol transmission, either from the patient or, much less likely, from the bedding the other nurse changed while she was present.

Once we'd surveyed and gotten samples from everyone, there was nothing for us to do but wait and turn our attention to the sick pathologist, whose name was Donato Aguilar. We figured if we tended to him ourselves, in addition to being able to help him we'd learn more about the disease and possibly keep a lid on the panic.

When Donato developed a fever, they closed Seton Hospital, which was a moot action since no one was going there anymore anyway. A Catholic hospital, Seton had a section reserved as a nuns' residence. When Donato got sick, most of the nuns moved out of there, except three or four who'd had close contact with him or the other patients. He was moved into a private room in the nuns' wing and two of the nuns were taking care of him. His was a high-profile case and several prestigious Bolivian physicians had visited the hospital to check on him, although no one wanted to actually risk going in to see him. They just asked the nuns for updates. Bob LeBow, who knew him personally through professional contact, had been the only physician to go in and examine him.

One neurologist in particular, president of the medical society in Cochabamba, proved a problem, nagging the nuns and looking down his nose at everybody. At one point, I was trying to explain some treatment issues to him in Spanish and he interrupted me to turn to an assistant to say, "This fellow doesn't understand how to use the subjunctive properly." I couldn't believe it. Here one of his colleagues is in the next room dying and he's criticizing my grammar! Our differences of opinion soon went beyond the use of the subjunctive.

As is too often the case, the source of the problems between us came down to politics. At that time, more so than today, certain areas in South America looked to Spain or France for medical training and had little shrift for a North American. Bolivia turned mostly to France. This guy was French-trained and wanted no part of my presumably inferior experience and training. The nuns, although primarily Bolivian and Colombian themselves, had a more favorable opinion of American medicine. They also saw that LeBow and I were actually going in to spend time with the patient. The nuns developed a very innovative approach to patient care: they created duplicate charts for Donato Aguilar.

When you're taking care of a patient, the chart is everything. It is a written record of the patient's vital statistics (temperature, pulse, respiration, blood pressure), all the laboratory data, hour-by-hour nursing notes, and the order page, where physicians write down exactly what they want done: what medication to give when, when to check vitals, everything. The nurses use this as a guide, checking off actions as they go.

For Donato, the nuns kept one for show, filled out, delivered to the Bolivian neurologist and his colleagues, and essentially ignored; the other was kept in a desk drawer, with orders by Bob and me that were followed. We never told the nuns to do this; they saw a problem and acted in what they felt to be in the best interest of their patient, in a totally nonconfrontational way, eliminating any need for a battle over the sick man's care. This was lucky for us because it would have been completely inappropriate for me or LeBow to outwardly over-rule any decision of the neurologist, since we were technically not even licensed to practice medicine in Bolivia. With the nun's complicity, though, we treated Donato and kept track of his illness ourselves.

Bob and I took turns caring for him, using mask, gown, and gloves. We would talk over Donato's case and the general state of affairs, making sure we had regular periodic samples of his blood frozen in liquid nitrogen for future study. It would be a shame if we lost any scientific information that could help the next cases. And then I had some epidemiological duties. It was necessary to check on close contacts, such as Donato's wife, and see if they had any signs of disease, continue the process of questioning and cross-questioning people who were involved, to try to pin down the exact facts surrounding the cases. And finally, we needed to do a serological study of the many hospital contacts who had no signs of disease. This would be a critical piece of information when we finally assessed how many people were actually infected from the index case.

Mercado, in the meantime, supervised the effort to find out where and how the epidemic started, overseeing the hunt for infected *Calomys* where the index case had stayed. I was grateful there were two of us back at the hospital; we each had time to rest, but Donato was still assured of round-the-clock attention by a physician.

When I first saw Donato, he was conscious and oriented but febrile and nervous as hell. He had severe muscle aches and his flesh was crawling with supersensitivity which made it uncomfortable for him even to feel the sheets against his skin, which is typical of South American hemorrhagic fevers. If it's true that doctors make the worst patients in general, you can just imagine how this guy felt. He'd seen firsthand the death of a friend with the disease and was inadvertently

responsible for his own infection. I'm not saying this made him difficult for us to deal with—I think he was tremendously appreciative
that we were there with him—but his illness seemed definitely harder
on him than it would have been for the average resident of Cochabamba.

As Aguilar's condition worsened, the situation grew more and
more surreal. It was as if we were holding our collective breath,
grateful at the end of every day that took us further beyond an
incubation period without a new case. The environment inside the
hospital was very tense. I always joke that this is one business where
you can usually get a table to yourself at the cafeteria wherever you
are, and this was certainly the case in Cochabamba.

Except for the few folks at the hospital, I had no contact with
anyone. It wasn't as if I could just go back to the hotel at the end of
my shift and call up Norma to decompress. Even to get through to
Karl and the others at MARU I'd have to patch together one or more
ham radios to reach Panama, then get a telephone link to the lab.
And that wasn't a great connection, just barely adequate for emergency circumstances, as well as being audible to every ham operator
tuned to that frequency.

So Bob and I focused on taking turns caring for Donato, leading a
double life with the nuns as far as the patient's chart was concerned,
while Mercado seemed to be having no luck in his search for answers
in the index case's hometown. They weren't finding many *Calomys*—
infected or not—and no one interviewed could shed any light on
where the index case might have been exposed. At the same time,
things weren't getting any better with Donato. He fell into a coma
(which was probably a blessing, considering the other things that
started going wrong for him physically), began bleeding more, then
went into shock.

About this time, a young woman came up to me during one of my
shifts and introduced herself as Donato's wife. A few years younger
than he was, she was strikingly beautiful and only just beginning to
show her pregnancy. "Oh, Doctor, I've got to talk to you. I've got to
talk to you," she repeated emphatically as she dragged me off into a
room where we could speak privately. It seemed that after Donato
cut himself doing the autopsy on the nurse, he didn't tell his wife
what had happened. He was aware of the presumed sexual transmission of these diseases—although not aware that it usually happened
in the convalescent part of the illness and not during incubation—so
he refused to have sex with her and tried to avoid even kissing her

good night, without explaining why. She grew afraid that perhaps he'd found another woman, or maybe the pregnancy turned him off. Whatever the reason, she was convinced he didn't love her anymore. Donato, no doubt, didn't want to alarm his wife and kept silent as he tried to do the manly thing and protect her and their unborn child. As for his wife, she did what any other young Latin American wife would do in her situation: she set out to win back her husband's love.

Poignantly, she described for me how she waited for him to come home one night and met him at the door in her best flimsy, see-through black nightgown. In this sexy costume, she did a little dance for him. One thing led to another and they finally had sex after several days of abstinence.

The next day, he developed a fever.

I tried to be both reassuring and realistic as I attempted to calm her. There was only the smallest of chances that she and the baby were exposed (although we really knew nothing about this disease). Also, I had to gently prepare her for what might happen with her husband.

From that day on, I became Mrs. Aguilar's confidant, which got harder and harder as his condition deteriorated. I had learned from my own mother's illness how these situations affect the patient's family and I tried to apply compassion to every case I treated, but I confess this case was more piercing to me than most. The Aguilars were young, in love, and happily expecting their first child before any of this started. We all make mistakes in our lives, have moments when we get careless. Donato Aguilar—by all accounts an exemplary physician—was paying dearly for his.

What is both scary and rewarding about working hemorrhagic fever cases is that most of the time you're dealing with young, productive lives. These patients are not eighty-five-year-old geriatrics with enlarged flabby hearts barely able to beat, already close to death's door. As sick as he was, I could tell that Donato Aguilar was young, muscular, and had been in terrific shape. With the work he did, I knew if we could pull him out of this we'd be making a difference to someone who had something to contribute in his lifetime, someone who truly had his whole life ahead of him—and that of his unborn child.

One of the most heartrending experiences I faced during the whole outbreak, including my conversations with Mrs. Aguilar, was talking to a dear friend of Donato's. The young man, a urologist, had been raised with Donato. They'd gone through school and medical school together. When the nurse died, this fellow volunteered to help him with the postmortem. "Donato, I'll come and help you do it. I don't want you to do this alone." The autopsy would be difficult given the dangerous risk of infection, but the nurse had been a friend of both of them and the urologist did not want Donato to be alone as he went through the emotional experience of cutting up somebody he'd known, liked, and worked with.

Together, they began the autopsy only three hours after the nurse died, trying to get her out of the hospital as soon as possible to limit exposure to others. Opening up the abdominal cavity, they were ready to do what we call "the pluck." You make a Y-shaped incision, open the chest and abdomen, then start at the mesentery, cutting away the attachments until you're able to pull out the organs. You usually do it in two parts—the thoracic organs first and then the abdominal organs.

Donato said, "Here, let me retract the small bowel and you trim away the root of the mesentery." The abdominal cavity is always full—there are coils of small intestine and the great loop of the large intestine. Since you open the body at the front, the only way to get around to the back next to the spine, to find the root of the mesentery anchoring the bowel, is to pull the organs away and to the side. Donato put his hands inside the abdominal cavity and pulled the bowel back toward him so his friend could cut away at the attachments.

But they made a crucial mistake. They used a scalpel. The urologist began trimming away at the root of the mesentery and it was he who nicked Aguilar's thumb, slicing right through the surgical rubber glove. Although Donato immediately pulled his finger back, squeezing out blood and quickly sticking it in the formalin they had ready for the organs they were going to preserve, this was direct blood-to-blood exposure. The formalin sterilized the cut, and we know that for some viruses like rabies extensive washing of a wound can prevent infection in experimental animals, so it was a good try. Donato scrubbed the cut, put on fresh gloves, and continued the autopsy, frightened but hoping for the best.

Instead, four days after the cut he was febrile, even though the usual incubation period is seven to fourteen days for BHF, and had been for the other cases in this outbreak. Why did Donato get so sick so much faster? Work with animals (and infections in humans) has shown that a larger dose of virus results in a shortened incubation period and a more dramatic course of illness. A bigger dose also increases the likelihood of mortality. At the time of the autopsy, the nurse was so viremic that even a small drop of her blood contained a large dose of virus particles. That scalpel cut exposed Donato to a lot more virus than others were exposed to, directly into his body. In essence, he mainlined virus.

And his illness was dramatic. As he continued to worsen, the urologist could only watch as his best friend suffered and neared a death he felt was brought about by his hands. For me, facing him was worse than talking to Donato's wife. He was another professional. He'd seen the nurse die and now he was watching his friend go downhill. There was nothing for me to say.

The course of Donato's illness was consistent with the other cases for which we had information. In the first few days they were feverish, with chills, malaise, weakness, and myalgia. The symptoms became gastrointestinal as the patient started feeling nauseated, began vomiting, developed diarrhea, experienced anorexia and abdominal tenderness in some cases. The bleeding tended to start about a week later.

By the seventh or eighth day, Donato was bleeding from just about everywhere there was to bleed from. Still comatose, he had bloody stool, blood in his urine, hemorrhaging around his gums, ecchymoses (under-the-skin bleeding) around the sites of all needle punctures, and petechial hemorrhages (smaller sites of discrete, in-the-skin bleeding) all over his body. He was in shock and his blood pressure was plummeting. We set him up on a drip of drugs to raise his blood pressure but we weren't sure how helpful it would be.

Aldo Paz, one of the hematologists at the hospital, offered to do blood-clotting studies and other blood work. We discussed the risk of exposure, but he felt if he was careful he'd be all right since there were no cases of infection among lab personnel. We appreciated his help, although all we really got out of the tests was confirmation that Donato had a severe coagulation defect, which we'd already observed clinically. But the test results underscored one of our difficulties in treating him: even if we'd had the means to test for antibodies to Machupo virus, which we didn't, it was too early for him to have developed them; and even if we'd been able to confirm he had Machupo virus, causing BHF, which we couldn't, there was no known treatment.

Donato started to have seizures. Everywhere we stuck him to take a sample he oozed blood. He began to go into respiratory failure. Bob and I were facing a classic dilemma: we'd given him a blood transfusion with Aldo Paz's help, and tried to push some fluids on him to bring his blood pressure up but it went into his lungs, giving him pulmonary edema. Today's potent diuretics were not available in Cochabamba, so we were stuck. We realized we had to ventilate this guy quickly.

On television shows it always looks easy to intubate a patient. In reality, it's not all that simple—even when the patient is not teeming with a deadly hemorrhagic fever virus. You have to get the head back in exactly the right position, hyperextending the neck, positioning the laryngoscope just so. If you're lucky then, you're looking right down the windpipe. If you're slightly out of practice, as LeBow and I were, you're probably looking at the base of the tongue. And with the agitation and cough that inevitably results from having a tube tickling the back of your throat and windpipe, we'd be dealing with the generation of a whole lot of aerosols—not an ideal situation, seeing that one nurse nearly died from watching a patient's linens get changed!

As we debated how to try this maneuver so as not to infect ourselves, one of the real heroes of the story stepped in. A doctor I'll call Dr. Ortega, an anesthesiologist who worked at the Social Security Hospital, stopped by to volunteer his help. "I've known Aguilar for years," he told us. "He is *muy amigo mio* [my good friend]. If you need anything, I'll help you with it. I don't care what the risk is. I work at the Social Security Hospital and I don't like those guys; they're all Communists anyway and Donato is my friend. You tell me if you need anything."

So we said, "Well, we need to do a cut-down and intubate him and it could be dangerous." We warned him about aerosols, the whole thing.

He replied, "If that's what has to be done, that's what has to be done."

He and I went in to intubate Donato wearing masks and gloves. I also had him put on a pair of clear glasses to protect his eyes, while I wore my regular glasses. I set up the respirator while he successfully intubated the patient. At last I was able to hook Donato up to the respirator. For a little while the ventilation seemed to help; gradually, however, Donato continued to drift downhill.

Bob and I decided not to try to resuscitate him when he finally crashed. It wasn't likely to help bring him back, it would put him through even more suffering, and it was just too risky. A short time later it was my job to tell Mrs. Aguilar that her husband was dead.

To this day, I've never had another experience quite like Donato's death. Every seasoned field epidemiologist I've talked about this with has had one case in an epidemic that becomes a defining moment in his life. There was probably no time from the moment I arrived there that I thought Donato was likely to recover or when I saw a good reason to hope. But he was so young and strong. If anyone could beat a disease like this, he could. I had just the briefest of conversations with him before he lost consciousness, but for days my single focus had been keeping him alive. It's devastating when you finally have to give up and it left me with the same feeling of emptiness I had when I learned that my mother had died.

I also felt that yes, he got sick because of a stupid mistake, but he certainly didn't deserve to die for it. In many ways, I identified with him: young, maybe a little too enthusiastic or overzealous; trying to do the right thing in the interest of advancing science, working to understand this horrible disease; professional enough not to over-react when he got exposed, but calmly making every effort to clean the wound and then get back to business; getting ready to start a family. There's a joke in the virus laboratory that the first guy who isolates a new, dangerous virus in the lab and gets infected is a hero. The second guy who gets infected gets a demotion. Donato is one of many martyrs to science who gave their lives in places most people in the United States have never heard of trying to stop one of these diseases. His infection and death personified one of my biggest fears about hazardous virus work. It takes more than hood lines, space suits, and blowers to make a safe laboratory. You need people with the understanding and preparation and experience to conduct themselves effectively in the presence of these potential killers.

With his death it became my turn to cut open someone I'd come to think of as a valued associate, if not a dear friend. There was something darkly ironic about making preparations for doing a post on a pathologist, particularly one who died as a result of an accident that occurred while he was doing a post.

At this point, LeBow and I parted ways. Although he was willing to do almost anything to keep Donato alive, he wanted no part in the autopsy. Mercado, however, agreed to help. I was in no position to turn him down, so I pulled out respirators, masks, gloves, and caps for both of us.

A Peruvian physician with the Pan American Health Organization (PAHO), the World Health Organization's branch in the Americas, offered to assist. He'd arrived a few days earlier to observe the outbreak and hadn't played any role in Donato's care up to that point. We asked him to stand outside the room where we would be doing the post to assist in packing away the samples. This would be tremendously helpful but wouldn't add another person—and another level of risk—to the procedure itself.

In addition to the protective wear, we established procedures to get the job done but keep everyone safe. On top of the regular surgical gloves we wore a pair of heavier rubber gloves like those used for washing dishes. Mercado agreed to use blunt scissors, which wouldn't penetrate a glove on contact the way a scalpel would. Perhaps most important, we vowed ahead of time that we would not both be inside the body at the same time. If one of us was cutting, the other would be outside the cadaver. In the short time that we worked together, Mercado and I had formed a mutual professional respect for each other and had grown to like each other personally. We'd both met Donato's friend who'd made the fateful cut on Donato and neither of us wanted to look the other in the eyes after having made the same mistake.

We decided to do the autopsy right there in the patient's bed. The last thing we wanted to do was drag his bloody body through the hospital to the proper autopsy table, alarming everybody and breaking the isolation. We wanted to get it done right, but we also wanted to get it done quickly so he could be out the door and the virus— whatever it was—contained.

Donato was pronounced dead at 11:25 P.M. on Sunday, March 7, and we started the autopsy at 12:15 A.M. We took the liver and spleen out first and had them in liquid nitrogen by 12:45. Since we didn't have the facilities to test samples from any of the other cases (and we were unsure even of the quality of those samples), Donato was our best hope for isolating the virus that caused these people to die. It was therefore critical we get these samples properly. We had taken blood from him every day and put it away in liquid nitrogen, but the organs—the spleen in particular—are really the best places to look for these viruses. We didn't want to waste any time getting the organs preserved.

Donato's body was a mess. There were some signs that were actually not consistent with Bolivian hemorrhagic fever but were in line with what we'd been told about the other cases in this outbreak. For example, Donato's sclera (the whites of his eyes) were yellow; three of the cases, unlike typical BHF patients, had signs of jaundice from their illness. There were no fresh petechiae at the time of autopsy, but there were older, fading lesions around his armpits and extensive ecchymoses along with extensive hemorrhages around his gums. His mouth was caked with blood.

When we opened the pleural cavity to take the lungs out, we found bloodstained fluid everywhere, with petechiae on the surface of the lungs and the surface of the pleural cavity. One good sign was that the lungs were mostly their normal pink and they didn't appear to be fluid-filled. He had gotten rid of some of the fluids we'd given him to keep his blood pressure up. But there were several areas scattered throughout that were so solid and red they looked like liver, in a state called hepatization. Later, the microscopic exam showed these were areas of secondary bacterial infection. While it probably didn't kill him, it definitely complicated the hemorrhagic fever, contributing to an impossible situation for his bodily defenses.

We found the heart enlarged, which meant it was failing. And there were a lot of petechial hemorrhages on the rest of the internal organs. The stomach was enlarged and contained about a liter of foul-smelling, bloody, clotted fluid. Viruses don't come through your respirator, but odors do, which didn't help ease our stress or make an already depressing job any more pleasant. But as distasteful as aspects of it were, and as tired as we were, Mercado and I couldn't afford to switch over to autopilot. We were ever mindful of how Donato's body came to be the subject of this post.

We'd been unable to find Donato's notes from the autopsy he performed on the nurse. From a scientific standpoint, therefore, unless we learned from the samples Mercado and I took from Donato's own body, his death would be in vain.

Mercado and I were working in the hospital room, up to our armpits in blood, and the Peruvian doctor stood just outside the door. He was dressed just as we were except that he wore a surgical mask instead of a respirator. As we took a sample, we would place the material in a vial, close it by screwing the cap on tight (not the kind of cap that could fall or pop off), wipe down the outside with disinfectant, and call out to the Peruvian doctor, who would open the door a crack and hold open a condom, into which we would drop the vial. He would then tie the condom securely so there was a double barrier of rubber and glass to protect anyone retrieving the biohazardous samples from the liquid nitrogen later. In addition to the sanitary protection, the condom helped to keep the nitrogen from getting in the vial.

Then, because the vials were only an inch and a half (3–4 cm) long and the liquid-nitrogen tank is two feet deep (and at 195° below zero, you can't just stick your hand in a tank of liquid nitrogen and fish around for samples), we also dropped the tied condoms into a nylon stocking. The way the tanks are designed, there's a little lip at the top where you could rest the top of the stockings without getting them in the liquid nitrogen. Opening the tank, then, you'd simply pull the top of the stocking to retrieve the sample.

Obviously, we'd come prepared. You can't very well rely on the nuns alone, however helpful and well intentioned, to get the supplies we were working with. Still, try ordering a gross of condoms and two dozen nylon stockings from the U.S. government! Add that to the scotch I'd learned to bring with me everywhere as a gift to expedite local assistance and you had all the basics for at least one other profession I could think of, not officially sanctioned by the Public Health Service or the U.S. government. We had a very down-to-earth, practical scientific use for these items; it just took a while to get a supply officer sitting in Bethesda, Maryland, to grasp the fact, but we got everything except the scotch, which we covered from our per diem. Safe sex isn't the only means by which condoms help prevent the spread of disease.

Mercado and I finished up the autopsy at about 3 A.M. and then came upon a problem. It would be nearly impossible to dress Donato for burial in the state he was in, especially in bed rather than on a table. We were exhausted. We'd been up all day and all night, our backs ached from leaning over the bed and we were just whacked. If this happened today, I would have vetoed dressing him (and sewing him up for that matter, because it's an unnecessary risk, another use of a needle). It wasn't as though we were preparing him for an open-casket funeral. But I was younger and less assertive back then and I was very respectful of Mercado's and the Aguilar family's wishes. I was in their country and it was their custom to sew up and dress the dead and Mercado was accustomed to doing this, so we replaced inside the cavities whatever organs (or organ parts) we hadn't frozen or fixed in formalin for later analysis and sewed him up properly, using as much care as our exhausted minds and bodies could muster.

But we still had the problem of getting his clothes on him. It's really very difficult to lift a dead man, even when you're not exhausted and emotionally spent. As much as I appreciated all Mercado had contributed to the effort, I was about to suggest we give up when he showed me how to cut the clothes in two and put them around Donato's corpse. We didn't have a proper body bag for him, but the undertakers had brought a coffin with a rubber lining.

When I had the time and perspective to look back on that night, two things touched me. The first was the extent of devastation in Donato Aguilar's body before he finally died. If I took away anything from the experience in Cochabamba it's a sense of the strength of these viruses. This was a virus that could kill anybody. You can't see them go through a town or kill such fit, healthy people without developing a tremendous respect for them. Nobody's invincible. Donato was my contemporary. Losing him to this virus was a lesson in my own mortality.

The second was Mercado's simple act of dressing Donato's body. Nobody would ever have known if Donato went into the ground wrapped in a bloody sheet, but Mercado was determined to give his countryman this final bit of dignity.

When Mercado and I finished up, we headed out to meet the Peruvian doctor and call it a day. As we were all getting out of our scrubs, the Peruvian doctor noticed a small abrasion on his thumb. He'd been wearing a double set of surgical gloves, which I wiped down with disinfectant and filled with water to test their integrity. Both pairs of gloves held the water, which proved that they were intact. Mercado and I were greatly relieved.

But the Peruvian doctor panicked as he went from not remembering if he'd had the abrasion before to being sure it came from his work that night.

"Look," I said calmly, "I don't know whether you had the abrasion before or not, but you know you can get an abrasion without any exposure. And we have inflated your gloves and they are intact. If there are no leaks in the gloves, how could anything get through?" I reminded him that all he did was stand outside the door holding condoms, waiting for us to drop into them vials which had been wiped down with disinfectant and which he never handled. Since he stayed outside the room at all times and wore a mask, even his aerosol exposure was minimal.

Still, he was unconvinced. He got on the ham radio and contacted the Pan American Health Organization and announced, "I cut myself while I was helping with the autopsy and I want to be evacuated and taken to the United States for medical care." Well, first of all, it wasn't true. Second, with everybody and his uncle tuned in to the ham circuits linked to Cochabamba, he added another element to the continuing soap opera we'd been working so hard to put to rest. I've since seen that there's always one guy in every epidemic who panics or otherwise distracts people from the mission, and that's the one you want to try to identify and select out beforehand. At the time, though, this was one of my first lessons in controlling the externals. While you can't control things like Costa Rican rainy seasons or mud slides, you can control team selection—or you should work pretty damn hard to try. You obviously don't want a guy like that serving as a critical part of your team. That's why now, as director of Special Pathogens, when there's a mysterious outbreak I want to make the decisions about who can and can't go; and I select people I know and trust rather than whoever may want to go or may be urged on me for political reasons.

The Peruvian doctor flew back to the United States and proved to be completely healthy and nonexposed. A Pan American Health Organization employee who worked in D.C., he was an inconvenience and a public relations embarrassment.

Luckily for us, while his plea got a lot of play in the Bolivian press, there wasn't much interest in the outbreak worldwide. Now that I've seen what can happen when you're trying to battle a killer virus and a media frenzy at the same time, I realize how lucky we were in Cochabamba. One of the nice things about going someplace remote is that you just give it your best shot and you deal only with the local people. You don't have a horde of reporters, each of whom has to get his own story, talk to you personally, get the best footage, and the rest of it.

Another negative impact from the broadcast was that the panic level got so high it looked like I had better get the precious samples to Panama quick, before the Bolivian government shut down air travel in and out of Cochabamba. We'd passed the end of an incubation period with no new cases (Donato's wife and unborn baby were considered out of the woods at that point) and there were no active cases to treat, so it made sense to leave immediately.

As I made travel arrangements and got ready for a night of celebration, drinking with the locals, before shipping out, LeBow and I received an urgent call from Dr. Ortega, the anesthesiologist who had bravely intubated Donato in the last days of his friend's life. He reported that he had a fever, his face was flushed and his eyes red.

Oh shit, I thought. These were all nonspecific symptoms, but they were also all common in the early stages of a hemorrhagic fever.

Although Donato's death had hit me hard, when I heard this I was utterly devastated. This man had put his life on the line to help us care for his friend and now he could be facing the same fate as Donato.

Bob LeBow and I discussed options. We knew that if we stayed, we were at risk of never getting out of there, and that if either of us became infected, no one was likely to step into our sickroom to treat us. If we tried to smuggle Ortega out, we risked exposing other passengers on the way. In my already weary mind I kept flashing to him deftly inserting the tube into Donato's trachea, desperately willing his friend to breathe. "We've got to get out of here," LeBow insisted. "They're gonna seal the airport and stop the flights."

I did the only thing a responsible Public Health Service physician could do: I went to the going-away party held in my honor and drank as much as was humanly possible for someone still getting a bearing on the altitude. Like Scarlett O'Hara, I figured we'd make sense of it all tomorrow.

Hung over and not naturally a morning person, I nevertheless made it to Dr. Ortega's house by 6 A.M. the next day. I wanted to examine him and make a decision while still leaving myself enough time to make the flight. His wife answered the door smiling! It seems that overnight he had erupted with a herpes zoster rash and this was the cause of his symptoms the night before. Never before had I been so delighted by the sight of shingles.

Before leaving, I contacted Karl from Cochabamba on the ham radio patched through to MARU. Karl greeted the news of my return rather oddly. "Okay, when you get to La Paz, give me a call on a secure line."

In La Paz I learned why Karl sounded so strange earlier. It seemed that since I'd been hanging out in Bolivia with the virus, the folks in the Canal Zone government definitely didn't want me back, nor was the government of Panama terribly enthusiastic over my return. It took all of Karl's negotiating skills to get me back in the country at all, and only after he worked out an arrangement whereby I'd be driven through Panama City directly to the Canal Zone in a "Do not pass Go, do not collect $200" fashion, where I'd be effectively quarantined at MARU in isolation in the converted attic apartment until everyone was convinced I wasn't a threat to public health. I later learned that if I became ill, an "unused building in the Canal Zone military hospital" would be opened for me since the local military governor had polled some of the Gorgas Hospital docs and they had declined to care for me. A team from Walter Reed was willing to fly down if necessary.

Fortunately, Norma and the kids were much more open-minded about the risk I posed. And I have to admit that after all I'd been through up in the Bolivian highlands, a couple of weeks in isolation with visits only from loved ones wasn't much of a hardship.

As it turned out, I'd soon be off again, trying to unlock the secret of Donato's death, worrying about the changing face of the disease, wondering whether there was some factor—possibly genetic—with the index case that made her more susceptible, or whether this was a strain of virus we hoped against hope would never come along.

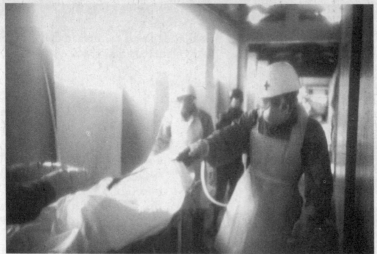

Field decontamination of body bags during the 1995 Kikwit, Zaire, Ebola outbreak. These are only a few of the 245 people known to have died in the epidemic.

Typical rural dwellings in the 1993 Four Corners hantavirus outbreak. Four people living here in this intersection of rodent habitat and human habitation developed hemorrhagic fever and two died.

Inoculating tissue cultures in the MARU lab, Panama, 1969.

My dad and me in Odessa, Texas, in the mid-1940s.

Karl Johnson, one of the great innovators in the field of modern infectious disease and my mentor at MARU, working in an early Level 4 lab setup at CDC, Atlanta.

Bolivia, 1993. There are many ways to hunt a virus. If you have to cross a savanna under two feet of water to get to the patients, horseback can be the most efficient.

Many factors can contribute to disease, as I've discovered throughout my career. This house, typical of the living conditions encountered throughout the Third World, provides a graphic example of why rodent- and arthropod-borne diseases, as well as common infections from poor water supply and waste disposal, are rampant.

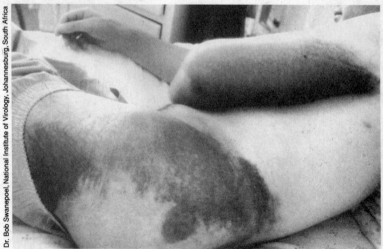

A typical hemorrhagic fever presentation—in this case the Crimean Congo virus. The five-dollar word for such extensive bleeding into the skin is "ecchymosis."

CDC

Working in the Level 4 lab takes an uncommon combination of skill, experience, and nerve. Here a laminar flow hood provides an extra level of protection in addition to the "space suit."

Author's collection

Egypt, 1978. Examining mosquito breeding in the Nile Delta during a Rift Valley fever epidemic with Charlie Bailey and Sherif el Said.

Stalking Ebola in the Central African Republic with Jean-Paul Gonzales and Jean-Paul Cornet. Despite the rigors of the expedition, there were some compensations, including full French meals with wine.

Me in front of Kitum Cave on the slope of Mount Elgon, Kenya. Though we've never been able to verify it, this is still on the long list of possible hiding places of Marburg.

The Gambia, 1987. One of the most important challenges in any fieldwork is explaining your mission to the local population and trying to secure their informed cooperation.

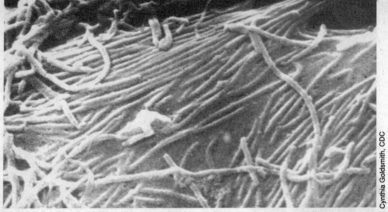

Ebola virus particles budding along the cell surface as they appear in three dimensions, magnified ten thousand times by the scanning electron microscope.

The USAMRIID team enters the Reston, Virginia, monkey house in the largest field biohazard operation we had ever faced.

A rare shot of me in uniform, with all my costume jewelry in place, at the end of my term at USAMRIID (1991). With me is the military artist who executed the C. J. Peters Award.

With my two daughters, Mayra and Antígona, at Mayra's high school graduation in San Diego, 1990.

San Ramón, Bolivia, 1993. Jorge Salazar and I finish up processing rodents during a hemorrhagic fever investigation.

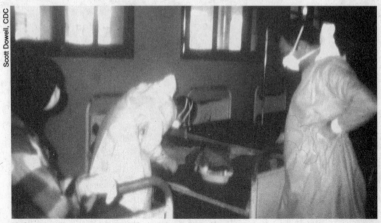

Kikwit, Zaire, 1995. When Pierre Rollin arrived at ground zero of the Ebola outbreak, he and his CDC colleagues had two goals: work with the international team to contain the virus and find out everything they could about Ebola.

My wife Susan and I in 1996, Rice University Distinguished Alumnus award.

In my office at CDC, my normal working clothes having replaced my military uniform.

5

More Answers,
More Questions

THE VIRUS that causes Bolivian hemorrhagic fever is called Machupo, named for the Machupo River, nourished by the rains that fall on the eastern slopes of the Andes where Cochabamba is situated. The rainwater floods the savannas of the Beni, where it is collected by the Machupo River as it flows across the Brazilian border to feed the Amazon. It's said that if you dip your hand into a river and then put it in again a few seconds later, you're putting your hand into a different river. Viruses are very much the same. The samples you've got and the symptomatology they represent may look the same as what you've seen in the far or recent past, but once you "dip your hand in"—well gloved and within the appropriate biosafety environment—you may find that you're dealing with a different organism and must react accordingly.

Given the different symptomatology of the patients in the Cochabamba outbreak, and the fact that it took place outside the endemic area of the Beni, we didn't know whether these patients actually had BHF. Consequently, we didn't know whether our samples contained Machupo virus or something else that was just as pathogenic and dangerous for man, if not more so. This dilemma had us running around in intellectual circles. If it was Machupo, we could do the tests safely in the Machupo lab at MARU, but we wouldn't know whether it was Machupo until we'd done the tests!

Karl was very sensitive to the risks involved in analyzing the samples at MARU. He'd had BHF and infected his significant other. The lab had already had one fatal exposure when Einar, a Bolivian technician, died. Each death in our business is a horrible one, but it would be rather ingenuous not to admit that patient deaths out in the field can often be regarded as statistics, while those that strike so close to home can never be just that. And on a very practical level, Karl couldn't afford to risk another infection—the human tragedy and the political fallout would have been too much. For these reasons, Karl thought it best to work up the samples in other, more secure facilities. Ever well connected, he contacted the folks who had the best biosafety setup going: the United States Army Medical Research Institute of Infectious Diseases at Fort Detrick in Frederick, Maryland. The gold standard for characterizing a virus is to isolate it—that is, to grow it from a sample in the lab. We had the personnel and we had the samples. All we needed was a good, safe lab.

In 1971, USAMRIID had the only true Biosafety Level 4 facilities in the United States. It also had an interesting history.

Up until 1968, Fort Detrick had been the center of the army's biological warfare program. It had an official and not very well-publicized name, but it was mainly referred to simply as "behind the fence," since the facilities were cut off from the rest of the fort's military tenants. For instance, Building 470 was a tall brick tower which could be seen from much of the surrounding countryside. What observers probably did not realize, however, was that it housed the military anthrax pilot plant, fermenting 2,500 gallons of *Bacillus anthracis* at a time, a hearty bacterium so efficient in its killing power that inhaling a high enough concentration of its wind-borne spores can lead to the deaths of thousands or tens of thousands of enemy soldiers. The other red-brick buildings scattered around the grassy Maryland turf contained all the equipment needed to convert a starter culture of anthrax, VEE, or tularemia into an efficient and effective biological weapon.

Chemical and biological warfare (CBW) had been proscribed by the 1925 Geneva Convention. That alone had not given the major combatants in World War I sufficient peace of mind to forget about the CBW threat. Some countries, like the United States, never ratified the convention, and most of the rest still maintained offensive CBW capabilities just in case the other side "used it first." The area behind the fence at Fort Detrick was our BW facility.

Despite concerns from a number of quarters, the U.S. offensive BW program was halted by President Richard Nixon in 1968, probably for considerations of both domestic and international image as much as anything else. We wanted to be perceived as taking the high road, being on the side of the angels.

Our change in public policy, of course, did not imply that everyone else—the Soviet Union, in particular, in those days—would follow suit. The generals may not have liked chemical and biological warfare, but if one was pinned to the wall, using CBW was preferable to losing, so it was reasonable to think that everyone would hedge their bets, especially if the other guy still had it. Therefore, it was decided that some kind of modern and highly capable defensive facility needed to be established—hence the United States Army Research Institute of Infectious Diseases or USAMRIID. It should also be noted that USAMRIID wasn't merely the old program with a new name. It was a new department with a new building in a new location "in front of the fence," with largely new staff so that it would be completely untainted by what Detrick had been known for in the past.

You might well ask why the military would need such a facility; after all, a lot of people and organizations—both within the government and without—work on vaccines and other agents for combating disease. I've spent most of my professional career in either the Public Health Service or the army, and I learned at the outset that public health medicine and military medicine each has its own distinct mission. Nobody would want the PHS to stint on AIDS research to work on a biological warfare threat such as anthrax, and the military wanted to have a group responsive to its needs and deadlines. In spite of this, the PHS and the army sometimes do similar things. And one thing I've tried to stress throughout my tenure in the army is that unless we can use our medical defenses in the field in the study of natural outbreaks—things like Venezuelan equine encephalitis and Machupo—unless we can show that what we've developed in the way of diagnostics and vaccines and antisera are useful in the field under real-life circumstances, they are not truly useful to the military. By that I mean that the army doesn't need vaccines on its shelves that have never been field-tested and may or may not work. The army needs vaccines that actually have been forged on the anvil of experience and can really be counted on not to give a false sense of security on the one hand and not to make more people sick than they protect on the other. We're not going to purposely infect soldiers or prisoners with weaponized diseases, as the Japanese Unit 731 did during World War II. But when there is an opportunity to study such diseases and ways to combat them in the real world, this is something military medicine is, and must be, uniquely set up to do responsibly and openly.

As I was to learn, there are things you can get done in the government that you can't get done anywhere else, and there are things that are routinely done everywhere else that simply can't be done in the government. Distinguishing between the two and reacting accordingly is the hallmark of the people—from yellow fever researcher Walter Reed to nuclear navy pioneer Hyman Rickover—who've left indelible impressions on military science and medicine.

The USAMRIID labs were shiny, squeaky-clean, and just waiting for a project to get their nonclassified program kick-started and provide justification for the dollars spent building the new place. Karl's proposition, then, was the ideal trade-off. Since we weren't sure our samples were really Machupo virus, we couldn't afford the risk of studying them at MARU. The army had state-of-the-art facilities with a top-notch support staff and nothing to look at as yet. And both MARU and the army would certainly reap benefits from any joint effort that showed cooperation between the Department of Defense and the Public Health Service.

From a selfish standpoint, the whole setup was perfect for me. It meant that I would get to be the first to use these laboratories—unlike anything I'd ever seen—and I would be the one to try to isolate and characterize the virus. Moreover, I had seen Donato's death, and I was determined to have that virus in a vial to understand its secrets.

I'm afraid my still newlywed but recently neglected wife, Norma, was a bit less happy with my new assignment than I was. I'd already left her and the kids just a few months after our wedding to head off into the middle of a deadly epidemic. Communication while I was in Cochabamba had ranged from extremely difficult to impossible. Upon my return, our personal life was inconvenienced by the fact that I had to live upstairs at the lab in semi-isolation for two weeks. Finally, just when Panamanian and Canal Zone authorities decided there would be no danger in returning me to the streets, I was home only long enough to pack for an indefinite assignment at a faraway lab in the United States. Norma and I were both learning an important lesson about the nature of the public health business: flexibility and understanding are key in any personal relationships.

After Karl wrapped up negotiations, I kissed Norma and the kids goodbye and packed up my samples. Although you could never do this today, I put the samples in a liquid-nitrogen tank small enough to carry on the plane and boarded a Braniff flight to Miami. The kindly flight attendant offered to help me with my unusual carry-on as I entered the cabin. "I have these biological specimens with me . . ." I started to explain.

"Oh, no problem," she responded cheerfully. "Let me put that here with the coats." The samples were well packed and well preserved, with multiple barriers between them and the people in the plane, but even so it wouldn't be permitted today.

Switching planes in Miami, I stowed the liquid-nitrogen tank between my knees. As we were preparing for takeoff, a flight attendant and the copilot came back to make sure my odd-looking device was safe. They were primarily concerned with the contents being under pressure, so I removed the exterior cap and showed them. Even with the top off, the samples were carefully frozen and sealed in the closed, screw-cap vials which had been wiped down with disinfectant, placed in condoms and in panty hose, and finally in the liquid nitrogen. I put the top back on my tank and we took off without incident.

When I arrived at National Airport in Washington, D.C., I was met by Ralph Kuehne, who was one of the senior technicians at USAMRIID, and his wife, Merle. Ralph had been assigned to be my shepherd while I was at USAMRIID, helping me in trying to isolate the virus. He and Merle also conveniently had a room I could rent while I was in Frederick.

As we waited for the rest of my bags, an eager redcap ran up and grabbed the liquid-nitrogen tank, turning it over on its side to place it on a dolly. Although the samples were safely tucked away behind protective barriers, there wasn't much between us and the liquid nitrogen. As this guy tilted the tank, before I could stop him some liquid nitrogen spilled out onto the floor and boiled away in a mist. The redcap's eyes suddenly grew about as big as the top of the canister as he watched the liquid vaporize instantly into a cloud. He stepped back smartly and let me take care of things from there—including the rest of my much heavier baggage.

As soon as we got outside I was assaulted by heat and humidity. Texas is hot, but it's a dry heat. Even the tropics of South America had an occasional breeze. I could not believe summertime in Washington, D.C. By comparison, the cooling breezes off the Pacific made Panama City positively balmy.

The accommodations at Ralph and Merle's house were a definite step up from living above the lab at MARU. Their son had recently left home, leaving a basement room with its own entrance, quite private and separate from the rest of the house. I found the accommodations and the company altogether pleasant and over the course of my stay we grew to be quite good friends.

The morning after my arrival, Ralph drove me to Detrick to begin work. It looked much like other army posts, with an exterior fence topped with barbed wire to discourage trespassers. Although the old biological warfare program was finished, many of the extra security precautions were still in place and entry was restricted. That's since changed. The security building where I checked in is now a credit union.

Since the official end of the offensive program in 1968, most of the people who worked behind the fence had either retired or been RIFed (let go in line with reduction-in-force procedures). Some remained, however, to dismantle the old labs in secrecy. During my stay, I was never cleared to go behind the fence; it was strictly need-to-know.

The USAMRIID building still looked new and stood out in sharp contrast to the aging red-brick structures that made up most of the rest of the base. White and modern, low to the ground, it was attractive, if stark. The building was divided into office areas, laboratories, and two wings that comprised the medical division. Fluorescent lighting reflected off the walls and clean tiled floors.

In the lab areas, similarities to any place I'd worked in the States or Panama disappeared. It was almost industrial in feel. The walls were painted battleship gray and there was a continuous hum from blowers that maintained proper direction of airflow throughout. The noise increased the farther back you went, and all the doors were closed to aid in compartmentalization into the "hot suites" where work with infectious agents was carried out. That section of the building had been symmetrically divided into four units and each unit into four hot suites, yielding sixteen individual "boxes," all designed to minimize exposure and cross-contamination.

Passing the hot suites, you entered a separate part of the building. Turning left, you arrived in the medical division, at a standard hospital nurses' station. Down the right side of the wing there was a long corridor that led to ordinary hospital rooms. Some of these had been converted temporarily into offices, but most of this area was used for experimental work with human volunteers. There was an active antimalaria drug program going on in which volunteers were injected with malarial parasites. The subjects were observed very closely and as soon as the parasites appeared in their blood they were treated with the new drugs. Their condition would be monitored and they would stay in these rooms for the entire duration of the test.

Down the left side of the medical division wing was another corridor which could be closed off and used for special studies of vaccines, so that volunteers would be protected from common infections. We could also change the airflow to make it negative pressure and care for patients infected with dangerous organisms. Directly in the center sat the "slammer." The slammer was a two-room suite in which you could take care of a patient with anything from a case of the sniffles to the most contagious and dangerous diseases imaginable. The slammer was run at Biosafety Level 4, with nurses and doctors in full space suits. It was originally designed to handle cases of laboratory accidents, a returning traveler who caught something nasty overseas, or maybe a couple of sick GIs. While the slammer could handle most, if not all, acute-care needs, none of us ever forgot that treatment was always the facility's secondary priority. Containment was the first.

Since the idea was to be able to quarantine and monitor people who'd been exposed to dangerous agents even before they became infectious (and therefore had a chance to expose others), the slammer was fitted with locks on the outside of the door only—in case patients got antsy being stuck inside for the course of an incubation period. There was also a "crash door" that led to another BSL 4 area set up as a small operating room. I couldn't help but wish we'd had facilities like this in Cochabamba when we had to intubate Donato and later perform the post.

There were two medical beds in the slammer, each in a separate room, so that, theoretically, two patients with two different infectious diseases could be treated at the same time. In an emergency (a biological warfare attack or catastrophic epidemic, for example), the other, "regular" hospital rooms along the left corridor could also be sealed off and used for BSL 4 patient care, with the medical staff in space suits or hoods fed from the air lines. With a fortunate dearth of laboratory accidents, chemical or biological warfare, and massive epidemics, the left corridor was used for vaccine trials. To fulfill its purpose, however, the slammer had to be ready for immediate use at any time.

In the late 1960s, it was thought that large-scale biological warfare would be initiated through generating large, invisible aerosol clouds that carried microorganisms to their victims, allowing easy deposit in the lungs of the target population. Virtually all the organisms or toxins worked with at USAMRIID, therefore, were ones that were infectious by the aerosol route. With this as a premise, airflow in the lab area was critical. A tremendous amount of thought and engineering went into the design of the air system, utilizing technology and experience developed over the years from work behind the fence. The engineering concepts developed and proven at Fort Detrick are the basis for systems used in universities and hospitals all over the world to protect microbiologists.

The airflow throughout the building was directed inward, with greater pressure in the corridors than in the containment suites so that any microscopic airborne escapees would be blown back into the containment area and up through ducts which would filter any dangerous particles before the air left the building. There was built-in redundancy of safety and function—everything from backup power and blowers to the ability to seal off and repressurize whole sections.

We were going to be doing our work in hot suite AA-4—Animal Assessment, Room 4. In the grand plan, four suites were dedicated for animal assessment, four for virology, four for bacteriology, and so on. Over time, however, as virology work took over hot suite after hot suite, the letters became more of a location designation than an indicator of the type of work done in the suite.

All the BSL 4 labs were laid out to offer the user optimum flexibility. There are two ways to maximally protect somebody who's working with an aerosol-infectious virus. Crudely stated, one is to put the virus in a bag and the other is to put the virus hunter in a bag. The labs at USAMRIID offered both options.

The "people in the bag" approach uses flexible plastic suits that completely isolate the wearer from the outside environment. Even the air the worker breathes is piped into the suit. The big advantage to this approach is that you don't have to buy or fashion special lab equipment, such as a microscope which is in the hot zone but whose eyepieces are outside. The disadvantage, from a personal point of view, is that once you're in that space suit, you're in. Before getting into it, you undress and put on a hospital scrub suit. After you've started your work, if you need to go to the bathroom, for example, you have to go through the whole process of scrubbing out and re-entering again. In the suit, the flow of air is loud. This, plus the physical barrier of the suit itself, makes it difficult to talk to people in the lab with you (whose hearing is blocked by their own suits) or use the phone.

Working in the suit alters your behavior, from logistics like considering the amount of coffee you drink that morning to less tangible elements like the strange, slightly disoriented sensation you have from being in this completely contained, separate environment. To me, it's similar to walking in a familiar room in the dark. There's the same feeling that you're just not quite as well oriented as you normally are. On the other hand, this "cut off" feeling allows a high level of concentration and introspection. After years of experience using them personally and observing others at work, I find about three hours to be the maximum amount of time the average person can work like that productively. There are some people who can work six or eight hours a day, but it takes a special personality to stay focused and alert. You don't want to get clumsy.

There are also some people who just can't work in a suit at all. It's like scuba diving: some people can't get used to being enclosed. I tried doing a postmortem once with a visiting pathologist from Walter Reed. He made it through the shower area but couldn't put the top of the suit over his head. He wanted to do it, but he was just too claustrophobic. As with fieldwork, you can't predict how someone will react until they're in it. And, over time, most people tend to burn out and need a break from it.

The other approach is the hood line, which is a grouping of airtight stainless-steel cabinets with gloves attached to ports that reach inside the cabinets. To work in the hood line, you simply put on a pair of surgical gloves and reach into the gloves fixed to the ports, which are made of neoprene—the thick material used in wet suits. The hood line is maintained at negative air pressure so much higher than the pressure in the suite that you can see the neoprene gloves "reach out" on their own, as though invisible ghosts had put them on.

The biggest advantage to working with the hood line is that you can set up an experiment, pull your gloved hands out of the hood, dip them in disinfectant, and go off to the bathroom, make a phone call, have a cup of coffee, get some more tissue culture media, look up the existing literature, whatever. In my work, it's common to set something up and then have about a half-hour or hour incubation period before moving on to the next step. In a suit, you pretty much have to sit down and wait it out or find some fill-in task, because it's not worth the time to go out and come back in. But if you've done it in a hood line you can easily make productive use of that time. Part of the flexibility of the AA-4 lab at USAMRIID was that the hood line was actually connected to a suited lab, so materials could be passed back and forth.

After considering all our options, we decided that to isolate the Cochabamba virus we'd use a hood line. I realized that my work not only would christen the Biosafety Level 4 equipment and facilities but would lay the groundwork for future procedures at USAMRIID. Although they'd been doing a lot of research with hot agents as part of the old offensive program, there was no institutional memory in the new labs. Even if you talked to some of the people who transferred from behind the fence, there was little crossover, because the equipment and procedures were all so new.

Once the decision was made, Ralph and I set out to staff our effort. We were very fortunate in finding a fellow named George Frye, who'd lost his post behind the fence when the offensive program closed. George was a short, redheaded guy who was wound up tight underneath but was very calm on the surface. During the time we worked together he proved he could deal with any problems that arose without flying off the handle, and his compulsiveness was a big advantage. This combination made him an ideal colleague for the work we were doing.

We were not equally fortunate in finding a pathologist to work with us, however. Ralph and I had decided that we needed someone to perform necropsies on any lab animals that died. I interviewed the military pathologist we were assigned, and he told me, "I've done a lot of work in BSL 4. I've autopsied monkeys, hamsters, mice, you name it; this is a piece of cake!" He sat down to demonstrate his expertise (on a mouse infected with a virus that was not dangerous to man) using his scalpel, sharp-pointed scissors, and so on. I looked on in horror, flashing to Donato's autopsy of the nurse, and told him we wouldn't be using scalpels and sharp scissors in our labs.

"Oh," he said, "I've always used them." I asked him what those animals were infected with. "Well," he replied, "yellow fever and tularemia."

"Have you had yellow fever vaccine?" I asked. He confirmed he had. "And tularemia responds to tetracycline, right? You've had the tularemia vaccine?" Again he nodded yes.

"Well," I explained, "there is no vaccine for this virus. There's no antibiotic, there's no therapy, there's no nothing. The mortality rate in the epidemic was five out of six and the samples we're working on came from a pathologist who was using a scalpel and got cut."

He looked at me, surprised, and said, "Oh, well, okay." That was the last we heard from him. We never did find anyone willing to follow our new animal autopsy procedures, so Ralph and I ended up doing it ourselves, which I didn't mind from the risk viewpoint because our procedures had several built-in safeguards.

My particular area of concern was learning to work with animals, because it's a fact that animals are absolutely the most dangerous part of a BSL 4 operation. Unless you get a breach in your suit or gloves, you're protected from infection by aerosol. That suit, however, is like tissue paper against the sharp teeth of an angry, scared, and infected animal. Not only that, but the same gear that's designed to protect you against aerosols actually creates an element of hazard by decreasing your orientation and your sense of touch. We'd be facing a double hazard: to grow a virus in an animal you have to inject it, which means dealing with both a sharp needle and sharp teeth.

While I knew animal work would be our biggest challenge and threat, I also knew it could be critical to our ability to isolate and study the virus. After all, rodents played a significant role in the spread of the Machupo or Machupo-like viruses. Newborn animals are generally most sensitive to viruses, partly because they have a less mature immune system, but also because cells from these animals are more susceptible to virus growth.

We knew that, of all rodents, hamsters are uniquely sensitive to several of the American arenaviruses. We planned to use cell culture to grow the virus too, but we wanted to stack the odds of success in our favor. I generally try to use cell cultures whenever I can, but sometimes there's no substitute for animals to provide sensitive systems to detect and assay viruses.

Anyone who's ever had a pet hamster knows them as delightful furry companions who love to be played with and handled. But a fully mature female who's been raised in a colony, who's never had any contact with humans, acting as an aggressive, protective mother, is a completely different animal. These hamsters hiss and bare their teeth and bite and are generally disagreeable critters.

Nonetheless, we needed hamsters younger than five days old—so young they had to be kept with their mothers in what we called a "shoe-box cage"—about fourteen inches long by seven inches wide, with a wire-mesh top, solid sides, and wood shavings in the bottom.

To inoculate the animals, or examine them, we had to reach into the cage and somehow get one of momma's babies from her.

The mother was particularly dangerous after we'd inoculated the babies (one at a time, being very careful not to stick ourselves through our gloves with the tiny needle), because they'd probably given her the virus. At that point, not only was the environment contaminated, but if she was infected and bit us, it would be like injecting ourselves with a needle. We began wearing thick leather gloves on top of the neoprene and rubber ones to give us an added layer of protection.

Another tricky thing about hamsters is that they have an incredible nesting reflex. People with pet hamsters know that if they get loose, they'll probably be found hidden in a shoe in the back of a closet somewhere. They will stuff anything resembling food or bedding in their cheek pouches to be carried back to their nest. In a lab environment, when baby hamsters are returned to the mother's cage, she immediately puts them in her cheek pouches. The catch is: if the mothers get too upset, they have a disturbing tendency to eat their young.

We "scrubbed" our SOPs (standard operating procedures) to get rid of everything sharp and all the glass that could shatter into sharp pieces. The only thing sharper than a mother hamster's teeth is a modern, carefully honed, brand-new disposable needle. I wanted no part of a needle—literally or figuratively—so I got rid of all sharp needles and we got our supplier, Becton Dickinson, to make us needles without points. It seemed like a great idea, but as we began to use the blunt needles a serious flaw emerged: they required so much pushing force to inject or penetrate anything that the risk of slipping and jabbing ourselves was actually greater. So, in the end, we went back to sharp needles because they gave us more control.

Finally, I swapped out all utensils like toothed forceps and scissors, replacing them with smooth, blunt versions. Instead of using scalpels, we'd mince the tissue with the blunt scissors. Rarely, under very defined circumstances, we'd take in a pair of razor blades to slice and dice something into tiny pieces for electron microscopy.

The hood line was shaped like an "E." At each of the points there was a mechanism for passing things into and out of the line. At one end we had something called a pass-through box, which is a sealed box with a very strong ultraviolet light. We kept it empty and clean so that whenever you needed to pass something in you could simply open the outer door, put the material inside, close the outer door, and open the inner door to move it into the hood line. After the inner door was opened, the pass-through box was potentially contaminated, so we'd spray it with disinfectant and let it sit with the UV light for fifteen minutes to decontaminate it. At the end of the middle branch of the "E" we had a double-door autoclave. We passed things in and out that way, running the autoclave to sterilize them (and the inside of the autoclave). And at the end of the last branch of the "E" was a dunk tank filled with disinfectant.

After weeks of practice, we finally had all the new policies and procedures in place and were ready to try the real thing. The very last step we took the weekend before we were to begin working with the samples from Cochabamba was to change the neoprene gloves in the hood. We'd heard from others at Fort Detrick that they had special gloves made with latex hands bonded onto neoprene sleeves. Since this was the government, a contract was prepared and the lowest bidder ginned up our gloves. We received these hybrid gloves, removed the usual neoprene gloves from the ports, and attached the new ones. This would give us a better sense of touch, and the latex wasn't nearly as slippery as the neoprene. These were not surgical gloves, however. The latex was thick enough to provide about the same protection as the neoprene.

The following Monday, we returned to work refreshed and ready to begin work on the samples. But I noticed the gloves were sagging a little bit—not a good sign. I examined them and found that the latex was detaching from the neoprene sleeves. Luckily for us we'd left them on over the weekend, so the leak occurred while we were setting up rather than while we were working up the samples from Bolivia. I bring up all of this mainly to impress upon you how crucial each step of the process may turn out to be and to show how a seemingly rational purchasing requirement (low bid) can complicate your life. To maintain biosafety with hot agents, you try to anticipate every possibility and leave nothing to chance. Lessons learned during the shakedown cruise of the USAMRIID lab would prove vitally important in the years to come.

Once we'd replaced the latex gloves with the original neoprene, we got the animals and the cell cultures in the hood line and started work with the samples. We had pieces of Donato's spleen, usually the best place in the body to find infectious virus particles in hemorrhagic fevers. We minced some of it into little pieces and ground it in a mortar (adding a small amount of sand to help with the grinding) until the material became a homogeneous slurry. We then centrifuged it to spin out all the big chunks and get a good dose of virus, since the virus particles are too small to sediment. Instead, they remain in the liquid at the top in what's called the supernatant. That material was what we used to try to grow—or isolate—the virus.

One of our key research methods was to inoculate all the hamsters and mice. Inoculating animals is the way viruses have been isolated from the beginning, going back to the turn of the century and Dr. Walter Reed's investigation of yellow fever, one of the classic tales of modern medicine.

We also added supernatant to a whole array of cell cultures—animal cells that have been adapted to grow artificially outside the animal.

But animals can be better for this work for several reasons. For one, a newborn mouse is cheap, easy to obtain, and fairly rugged. In remote areas of Africa or South America, you'll be able to raise mice but it can be virtually impossible to grow a cell culture. If you're one of the occasional readers who've used cell cultures in the United States and don't think it should be a problem, trust me on this: it is.

As I've indicated, the newborn mice and hamsters, with their highly susceptible cells and less competent immune systems, give the virus a head start. This ease of propagation is important because, ideally, you want to grow a virus in living cells through several generations so you can characterize it and fully examine its antigenicity. And practically speaking, it's usually easier to keep a mouse or hamster alive than it is to keep a cell culture growing. It's hard to keep a cell culture sterile; it requires close monitoring of the temperature in the incubators; the tissue culture medium has to be prepared very carefully and the water going into it has to be pure. Even then, assuming you're successful, many viruses give a much higher yield in the tissues of a single mouse than we can get from enormous quantities of cell culture. Some viruses cannot be grown in a cell culture at all—only a live animal will do.

With all these arguments for animal research, we still wanted to use cell cultures too, because if you have the infrastructure to support it (which we certainly did at USAMRIID), you have better control and the system is better characterized than in a complex living creature.

Even after getting a virus to grow, finding it can be tricky in either animals or cultures. Let's say you inoculate a mouse. You wait and watch for the mouse to get sick. It's possible the virus could infect the mouse, continue to grow, and yet the mouse would show no symptoms at all. Unless you test samples from the mouse you'll miss it.

In cell cultures, different types of viruses grow and reveal themselves in different ways. Some viruses, like the one that causes Rift Valley fever, are so destructive that within a period of twelve to forty-eight hours, depending on how much virus you added to your culture, all the cells could be dead. With other types of viruses—like hantaviruses and arenaviruses—there could be little or no observable damage. We call such damage cytopathic effect, or CPE, and it can be seen under a microscope.

Machupo is an arenavirus, so we knew that if the Cochabamba samples did contain Machupo, we wouldn't be able to rely on observable CPE to indicate it was there. As a result, we inoculated the cells, observed them for about a week or so visually, and then tested them for viral antigens. It took several rounds of testing samples from Donato and others from Cochabamba to confirm our findings, but eventually we isolated incredibly high titers of Machupo.

Titer is an important measurement in infectious disease because it shows us how powerful or dangerous a given preparation of an agent may be. If we take, for example, a virus preparation and dilute it, the highest dilution that's still positive for the virus—in other words, that's still potent in the assay used—is the titer. If you diluted some particular virus 10 to 1 and it was still effective, then you diluted it 100 to 1 and it was still effective, then you diluted it 1,000 to 1 and it wasn't effective any longer, 100 to 1 was the last effective point, so we'd say that the titer for that virus strain was 1 to 100. That means that in a dilution of 1 part in 100 this virus can still cause disease. The titers of Machupo virus in our samples could be diluted 1 to 1 million and still were positive, a dangerous liquid best kept within our BSL 4 laboratory.

When we actually found the evidence of Machupo in our samples, I felt a combination of exhilaration and letdown. This is not an unusual reaction in our line of work. On the one hand, it was exhilarating because it had been our first hypothesis, we kind of expected to find it, and we hadn't missed it. Our instincts had been right on target. Our samples had come from someone who died after a very fulminant illness. Some were taken from the spleen, which is usually the best organ in which to find virus, and they'd been preserved very quickly after the patient's death, so we fully expected to find *something*. But just as it was exciting finally to have a positive ID on the killer, there's always a little bit of disappointment that you haven't discovered something completely new. It also scares the crap out of you to think that you may have been rummaging around in Pandora's box. Nonetheless, if anyone comes up with it, you want to be the one to find it and neutralize it.

Please God, don't let this happen. But if it does, let it happen on my watch.

Once we had shown that the Cochabamba disease was caused by Machupo virus, we wanted to find out if it was a different strain. The index case had come from the Beni, but we hadn't been able to pinpoint a particular high-risk situation for exposure. She hadn't, for example, visited a house full of infected *Calomys* or spent time out harvesting bananas in a known dangerous area. On the contrary, after extensive trapping on the ranch she visited, only thirty rodents were caught and none of them were *Calomys callosus*, the only known natural reservoir for Machupo.

Absent such evidence, we had to assume that the woman had been to the region and somehow been exposed there. Sometimes you have to live with loose ends and question marks in fieldwork. I certainly hoped nature wouldn't repeat the experiment, but she probably will someday. And there were some signs that the disease was different. Three patients had developed jaundice, not part of the Machupo picture as we knew it but the high altitudes of Cochabamba possibly could have resulted in different clinical signs. We tried putting the Cochabamba virus and control Machupo samples from San Joaquín in mice and other animals and found several differences. The lethality of this virus for guinea pigs was slightly greater, for example, which could weigh against the hypothesis that it was the identical virus. But we simply didn't have the technology to prove it. As a result, we had to be satisfied with the fact that we'd isolated the virus.

Today, we would be able to analyze its genetic structure, determine the exact nucleotide sequence of the entire genome of this virus, and compare it with the prototype virus' genome. In fact, some viruses are actually simple enough so that we can use reverse transcriptase, a replicating enzyme for RNA genes, to construct a DNA clone of their entire RNA genome. But while we can spot differences in the genome today, that still won't tell us what those differences do. We can't just see that it's different and say, "Oh, that will cause jaundice," or render a deadly disease harmless, or aerosol-infectious, or whatever. We also don't have a really good model system to test how the difference will affect humans except by infecting humans (not a good choice, for all the obvious reasons) or monkeys (which are expensive and difficult to work with).

We did use the monkeys in other settings. At the time of the Reston Ebola outbreak in 1989, this method was effective in testing the difference between the Ebola strain we were finding in the monkey house and the strain that had devastated entire villages in Zaire.

There have been scattered cases of Machupo since then, though in the Beni rather than the mountains. There were no more diagnosed cases until 1975, and then not again until one turned up in 1993. But the next year there was a true outbreak in the town of Magdalena, where one man who worked out in the fields came down with the virus and exposed his entire family. Although he survived, all six family members infected, including his wife and children, died over a period of about three weeks. One of the cases in the family outbreak became ill with no real direct contact with the patient, leading us to suspect an aerosol-mediated infection as we had seen with the nurse in Cochabamba. Because of the remote, rural nature of much of the Beni, it's possible there have been more cases and they just weren't recognized or reported. But the disease itself is such a dramatic medical syndrome it's also likely that had there been a case of full-blown Bolivian hemorrhagic fever it would have been radioed in even from a remote area. They certainly recognized it in 1971—in a place it had never been seen before.

It's disturbing when you establish procedures such as rodent control to avoid a disease, and there's a drop in cases over a long period of time, and then suddenly the disease starts to crop up again. We can argue about reporting accuracy and speculate as to whether the virus has evolved in some way that makes it more infectious. But I think it's also just as likely there's been an increased number of cases in the 1990s because there are more infected *Calomys* in contact with humans, just as increased numbers of field mice in the southwestern United States increases the risk of hantavirus pulmonary syndrome. We all live on this planet together, man and mouse and everything in between.

While isolating a particular virus is important, just as important, in the big picture, is developing the specific skills and instincts and institutional competence to deal with potentially deadly agents in rigorous biosafety conditions. It is crucial that when the next crisis occurs—as it most certainly will at some point—you don't want to take time out to reinvent the wheel and you certainly don't want to divert your efforts because of a laboratory outbreak. You need to know what to do right away and whom to rely on.

On our "shakedown cruise" at USAMRIID, we had our share of human trials and errors. Even at Biosafety Level 4 a certain amount of routine sets in. I recall a sphincter-tightening episode one afternoon when George was ready to go to lunch and couldn't get the autoclave door to open, even though he was sure he'd run it. In the old days behind the fence, the only person who had a key to override the autoclave door-locking mechanism was the supervisor. Since we were all well-trained and enlightened adults, however, I'd tried to institute a more democratic policy. Everyone knew the key was kept in my desk and I trusted them to be able to use it when necessary. After all, I wasn't running a ministerial toilet in Costa Rica. Well, George went into my desk, got the key, defeated the interlock, opened the autoclave door, and the alarm sounded. He *hadn't* run the autoclave—and the inside autoclave door was open! The inside of the hood line was open to the room. Fortunately, he didn't take anything out and the negative air pressure in the hood sucked the air from the room strongly back in, so there wasn't any real exposure, but it could have been a very dangerous situation. We very quickly went back to the old policy of one person controlling the key.

That last situation drove home yet once more how critical human beings are in the biosafety equation. None of the equipment is worth a damn if you can't (or don't) follow safe practices. And you can't act as your own supervisor. Just as you want built-in redundancy in the power, blowers, and the rest, a certain amount of human redundancy has to be in place. The supervisor, in that case, would have played the role of human redundancy. George was sure of himself, sure he'd run the autoclave, but a second set of eyes should have caught the open door and forced him to run it all again to be on the safe side.

Although we went through our share of challenges in our search for this virus, the work itself was very rewarding, which was particularly important to me because this was a difficult period in my personal life. To begin with, I'd spent more time with Norma and the kids before we got married than I had after. It was too expensive to call more than occasionally, so our relationship consisted of sporadic letters and one visit. One of the guys at Fort Detrick went on vacation and let us stay in his house for the couple of weeks he was away. But the visit almost made things more difficult after they left for home without me.

Worse, even before their visit—in the middle of working up the samples from Cochabamba—I got an unexpected call from Odessa, Texas. My father had been in a terrible automobile accident. He was severely injured, in the intensive-care unit, and they didn't think he would make it. I immediately flew down to Odessa. When I got there I found out that he and his second wife were in their Ford when a pickup truck hit them smack on the driver's side. My dad suffered some crushing injuries to the chest and was in and out of consciousness, at times delirious and at other times comatose. We didn't know whether he had permanent brain damage because he didn't have enough oxygen in his blood, or was temporarily impaired. He was in the ICU for a couple of weeks. I'd gone back to Frederick when it seemed he might stabilize, because they needed me at the lab. But I ended up having to turn right around and go back to Odessa when I learned he had died.

My father's death was hard for me. When I left the hospital I knew the prognosis wasn't good, but we thought he would make it out of ICU. I knew it was possible that he might die, but I just wasn't ready emotionally. It was so different from my mother's death, which was more wrenching in some ways because I was so young and saw her disintegration every day. But at least we'd had time to prepare ourselves. Even though neither one of us wrote or called much, my dad and I had always been close, and with whatever dangers and uncertainties I faced in my career, I knew he was out there for me as an emotional anchor. The accident was so sudden and so unexpected that I never even had a chance to say goodbye.

I also never had a chance to introduce him to my new family. Norma and the kids had intended to meet him during their visit to the States but the accident happened just before their arrival. There was no continuity, no shared history, however brief, between the family I came from and the one I had become part of. I'd met my dad's wife before the accident, and while I liked her a great deal (she survived him by a few years), since I was an only child, my dad was the only close family I had left.

So despite the rewards of my work at Fort Detrick, I was glad to get back to Panama and home. Norma and I didn't waste any time getting our marriage back on track. She became pregnant shortly after my return, and our daughter Mayra was born October 27, 1972. Her birth helped to bring the loss of my parents full circle. Meanwhile, I was trying finally to wrap up the lab work from the hepatitis studies interrupted by the Cochabamba outbreak.

If there's been one pattern in my life, however, it's that just when I think things are finally settling down, something happens to shake it all up again, generally from a highly unexpected source. And the source this time, so to speak, was the President of the United States.

In 1972, President Richard Nixon declared war on cancer. This, in itself, was a very noble pronouncement. For years, many of us had been saying things such as "In a country as rich and powerful as ours, if we'd just spend as much on X as we do on Y . . . ," with X typically representing medical care or scientific research, and Y invariably representing something having to do with particularly stupid or wasteful programs, usually in the military-industrial complex. So here was one case where the President and the leaders of both political parties were saying, "Let's fight this dread disease that devastates so many of us and identify it as the enemy it really is."

But every war has its casualties. Soon enough, the announcement came from on high that the National Cancer Institute needed personnel slots from NIAID—the National Institute of Allergy and Infectious Diseases—and the decision was made to close the lab at MARU. It was one of those classic turf wars that happen when dollars and personnel get reshuffled in the government. I'm sure the folks at NIAID looked around and said, "Oh, we can close one of the labs here in Bethesda where my next-door neighbor—the guy I carpool with—works, or we can close the Rocky Mountain lab in Montana, but I'm sure as hell not going to be the one to break the news to Senator Mike Mansfield. Or we could do this place called the Middle America Research Unit in Panama. I know they were reviewed and told they were doing very good work, but we don't know anybody from there, and probably 99 percent of the American public doesn't know anything about Panama except that we operate a canal down there."

NIAID director Dr. John Seal announced the closure of MARU, stating that Public Health Service personnel affected by the shutdown would be given one year's support to get their careers back on track. This was critical to someone like me who didn't have any cushion to support my growing family.

The idea of one year's support, however, was of little consolation compared with the loss of the lab and my work and life in Panama. I'd been there long enough now so that in addition to my extended family through Norma, I had a lot of good friends. My Spanish had developed to the point where I could go to a party and converse intelligently (or at least so I thought) even with people I'd never met before.

All in all, I liked my work, loved the climate and the beauty of the country, and realized I would sorely miss the lifestyle.

Karl, however, was pissed off. This was the guy who'd first isolated the virus that causes Bolivian hemorrhagic fever, who, with Fred Murphy and Wallace Rowe, had defined the new virus family Arenaviridae, and made a major contribution to the nation's health with his Venezuelan equine encephalitis work in both Costa Rica and the United States, providing the scientific basis to stop it when it got to Texas. He was a major player in public health and now he was being told to pack it in.

Karl was too important a guy to just let go, so they extended his contract for a couple of years and merged his research work with what was going on at the Gorgas Memorial Laboratory in Panama. He stayed there until he went to CDC as chief of viral Special Pathogens (the job I now hold).

When we first heard the lab was to be closed, our associate Bob Tesh persuaded us to undertake a little campaign to save MARU. Bob wrote letters to various overseas contacts who could testify to the usefulness of the contribution we'd been making, and to influential members of Congress and other parts of the U.S. government. The letters were signed by folks at the lab, including me, pointing out the shortsightedness of closing the lab.

Being naive and inexperienced in the machinations of bureaucracy, I had no concept of the implications of my actions. I would soon find out. On one of my visits back to NIH to see Bob Purcell about my hepatitis research, John Seal called me into his office and when I closed the door behind me began to rant: "By God, all of tropical virology is dead and we're killing it! You can write all the letters you want but we don't have the money. We're gonna kill it and I don't care who you wrote letters to! If it gets any tighter, parasitology will be the next to go!"

Then came the punch line: "Furthermore, the offer that I gave you for a year's support after you leave here is dead too! Forget it. That's it, you're out of here!"

And that was that. Years later, I went to Hamilton, Montana, to interview for a job at the Rocky Mountain Lab and learned that Seal hadn't forgotten about me. The person interviewing me was named John Coe. We'd worked together on *Calomys* immunoglobulins in Panama and we were old friends. While I was there, the director called Bethesda to see where I stood in terms of my Public Health Service commission.

The first thing I heard him say into the telephone was "Hi, John." Then he said, "We've got a guy here that we're talking to about a job and I'd like to find out about his old PHS status and how we can get him back in if this goes any further. His name is Peters. . . . Yes, C. J. Peters. . . . Yeah, that's right, the one that used to be in Panama. Oh? Oh? Oh? Okay, thanks, John."

He hung up, turned to me, and said apologetically, "I don't know what you did to John Seal, but you're not going to work for NIAID ever again in any capacity as long as John's there."

I learned an important lesson about how bureaucracies function from that. You can't make someone higher up in the food chain look bad; this is a cardinal sin in government. You can say whatever you want that's critical of your organization as long as it benefits the organization, but if it ever brings down wrath upon the organization, you're in trouble. I'd broken the bureaucracy's oath of silence and John Seal was going to punish me.

While I felt chagrined, knowing the way I was in those days, even if I'd understood the consequences, I probably would have pursued the letter campaign—perhaps even more aggressively—because I thought the decision was wrong as well as unfair. Today I can look back and laugh, but it hurt to feel so powerless and disconnected. I realized I knew more about how to negotiate my way around in Latin America than I did in my own country.

Clearly, I had to move on. I knew the logical thing for me to do was to finish up my internal medicine training and take the boards. I wanted to stay in infectious disease, but fellowships were few and far between. As fruitful as research in Panama had been, you get to a certain point in your career where if you stray too far off the traditional track, you risk missing opportunities later, because people don't view you as being serious. I applied to and was accepted into a highly competitive residency program at the University of California at San Diego hospital, and that decided my course.

Norma and I took all the money we had and packed up the kids for the move to California. Despite her revolutionary background and inherent spirit of adventure, she found it difficult to leave her family, friends, and job and move so far away. We also had to leave behind a prized possession known as La Bala de Plata, the Silver Bullet, a mid-1960s silver Plymouth Barracuda convertible I'd bought used in Dallas and driven while I was in Panama. It was pretty much shot—mutant tropical cockroaches lived in the floor mats and the white top was covered with black fungus from the rain—but it represented a lot of good times.

Early in our courtship, Norma and I had discussed the fact that I would be in Panama only for a few years before I'd have to return to the States. When we talked about marriage, it was with the understanding that we'd move—as a family—to the States when my time was up. In retrospect, though, I realize that what you say when you're falling in love—when everything, including a prospective future in another country, feels exciting and romantic and new—is a lot different from actually living it.

In the beginning, the change was fun and stimulating for Norma, who really seemed to enjoy the cultural differences. As months passed, however, I think she started to miss her life in Panama. The biggest adjustment was that in Panama there was an entire class of women who cleaned and took care of children whom we could hire to stay at our house six days a week for a modest salary. We provided lodging, food, and about sixty or seventy dollars a month (no small amount by Panamanian standards back then). Now, cleaning the house was no big deal. After all, we had a vacuum cleaner and dishwasher and all the other modern American conveniences. But in Panama, with the help she had from these women and from her mother, Norma had a babysitter available whenever we wanted to get out of the house for a while. In San Diego, we certainly didn't have the money to hire a nanny and I was so busy with my residency that she had to stay home with the three kids (ages ten, nine, and not yet one) virtually 100 percent of the time.

Back in Panama, she'd also had a very active social and professional life outside the home. Norma came from a middle-class Panamanian family. Her father, who died before I met her, had been a prominent lawyer. She had graduated from college, her English was excellent, she had passed all the exams to be U.S.-accredited as a medical technician and she had a solid career at Gorgas Hospital. In California, she had no outside job of her own and missed both her former status and her network of family and friends. After life became more settled for us, I encouraged her to take a refresher course and get reaccredited so she could continue the work she'd loved in Panama, figuring she'd meet new people and get back in the swim of things, but my suggestions didn't seem to make much of an impression.

In some ways, it was similar to the time I moved to Texas with Lea.

A relationship that worked well in one context didn't necessarily translate to a new place, a new culture, a new set of social parameters. Still, I was hopeful that as my career settled down and as Norma gradually found her niche in San Diego, our lives together would become less stressful.

Even though I knew my ultimate goal was research, I threw myself 100 percent into clinical training, which has served me well, because ultimately, though we tend to forget it in the heat of an unexplained outbreak, the ultimate concern is the individual patient. When we talk about the clinical aspects of hemorrhagic fevers, what we're really talking about is the devastating effects of disease on individual human beings. As I approached the end of my year at UCSD and passed my boards in internal medicine, I wanted to get back to research on arenaviruses. In the United States, the big arenavirus is LCM—lymphocytic choriomeningitis virus.

The Scripps Clinic and Research Foundation in La Jolla, California, had an outstanding research team working on LCM at the time, headed by the world-renowned immunologist Frank Dixon. I wrote Dr. Dixon a "you don't know me but" letter from Panama, explaining my interest in chronic viral infections, outlining my experience and my hope that I could work with him for a couple of years researching LCM. He sent a brief letter back saying they'd be glad to have me once I was done with my year of clinical work and he'd arrange funding for my work. I was ecstatic; things were definitely looking up.

But the first day I showed up at Scripps, Dixon said, "Oh, you're here. Okay, good. Well, let's see. We need to get you a fellowship." He turned to his secretary and said, "Sally, get those Cancer Society forms out for this fellow and have him fill them out and we'll get that Cancer Society fellowship for him."

Sally reminded him that the number of fellows per mentor was limited to one and he already had one. He responded, "All right, get Dick Lerner in here and have him write it up." Dick Lerner is now head of Scripps and I didn't mind working under him if it meant I could do the research, but I thought I was there to work on the LCM virus. But it simply wasn't to be. The War on Cancer had shifted the entire focus of his lab just as it had forced the closure of MARU. I ended up working on leukemia viruses, studying tumor transplants and later autoimmune mice. Any disappointment I might have felt soon passed, though, when it became apparent that the switch in direction was leading me on a better path.

Scripps was among the handful of top immunology centers in the world, with leading people in every aspect of immunology on staff. Everybody who was anybody in the field came to Scripps at some point to find out what we were working on and give a seminar on their own research. It was an incredible vantage point from which to learn immunology.

My time there, too, was during the "good old days" of scientific research, when funding was still readily available. I worked under both Dixon and Lerner, but if I had another idea and wanted to work on it in my spare time, I could. There was enough money around so that you could explore new areas on the side, which could lead in new and unexpected directions. Today, grants are tight and so the creative aspect of science tends to suffer. It seems there is no room left for serendipity, one of the most potent of all forces in science. If penicillin pioneer Alexander Fleming were working in today's climate, I wonder if he would have just pitched out the culture dishes which had been ruined by some fungus, because that's not what he was looking for and his grant renewal was due. These days, funding is increasingly difficult to find, and when you have it, you still have to worry that the way you use your budget will have an impact on your ability to get the next grant. There's often a slippery tightrope on which you have to balance. If you propose something that's been done before, it's unchallenging and probably won't extend scientific knowledge; there's no reason to fund it. But if you propose anything too revolutionary or avant-garde, then the reviewers may not "get it." As a result, I find a lot of grant money is used for reinventing the wheel, at least up to a point, before getting on with what the researcher really wants to do. And I think we all suffer in terms of the speed of our progress.

There is virtually no such thing as a scientific researcher working alone and unfettered any longer. Almost everything takes money. If you need to use animals, you have to jump through many hoops having your protocols reviewed, too often by people who have little appreciation of what you're trying to accomplish or what the benefit to science will be. As anyone who works with me knows, I am as great an advocate of humane treatment of laboratory animals as any researcher, but we can't treat lab animals as pets; it just won't work. And there are many other examples of constraints on research today.

Thus, although the technology we have today is leaps and bounds ahead of anything we had twenty-five years ago, in some ways there was more freedom when I was in my residency days, allowing more people to make progress in more diverse areas.

If funding was relatively free at the lab, our finances at home were pinched. Norma and I had bought a fairly nice home in Clairmont, in the San Diego suburbs. With three kids and a mortgage, my income didn't go far. I was making a better salary as a postdoctoral fellow than I did as a resident at UCSD, but certainly not as much as I was making in Panama as a PHS officer, and we were really starting to miss Norma's income. I decided it would make things easier if I got another job, so I started moonlighting at the clinic at Scripps.

The clinic in La Jolla could best be described as specializing in the diseases of the rich. The regular docs there, while dedicated professionals, had no interest in working nights; they were only too happy to have folks like me fill in after dark. At first I earned just a flat fee for staying the night, but as the work picked up and the peons like me made more noise, we were allowed to bill by the hour. We still got paid a flat rate for the night, plus whatever we billed.

This made it a lot easier to deal with some of the patients. Although there was the occasional heart attack and other serious illnesses, there were also a lot of nuisance calls. One common complaint we'd see were the wealthy who came in with headaches. To be fair, Scripps specialized in headaches, and a lot of these patients lived in constant pain that affected every aspect of their lives. But there were also people who were just looking for a prescription for whatever they were addicted to.

I'd never worked in private practice before—and this wasn't just ordinary, run-of-the-mill private practice. This was the private practice to the rich. The economic contrast showed itself in everything from the symptoms presented and the common ailments described to the philosophical approach to medicine.

Norma and I were having more problems at home. Over time, it seemed we had different views on everything, from basic child-rearing values and handling personal finances to politics. You always think of there being one big catalyst or triggering event to break up a marriage: someone falls out of love, or in love with someone else. But for me it was just a lot of small but significant things that kept stacking up.

Norma wasn't happy either. At one point, after we'd had a fight about money—even with my working two jobs, we didn't have the bucks to send her and the kids back to Panama to visit her family— Norma announced she would get a job. I thought: *Great! This will give her an outlet; it could be really good for us.* But instead of helping, it simply meant more time apart from each other, allowing the other problems between us to fester.

I realized we were in real trouble after the office she worked at extended its hours to include Wednesday evenings. I had to go home early every Wednesday to take care of the kids and found that quickly became my favorite day of the week: a short workday, time to relax and play with the children, and no Norma until late. I knew then the writing was on the wall.

As 1976 drew to a close, I was nearing the end of my fellowship at Scripps and I found myself at a real crossroads. I enjoyed my work in immunology but I was starting to feel a pull to return to tropical viruses. That fall I attended an annual tropical medicine meeting and heard an audiotape made by Karl Johnson in Kinshasa, Zaire, at ground zero of an outbreak of a horrific new hemorrhagic virus. Karl was now chief of viral Special Pathogens at CDC, and the disease sounded like something out of *The Andromeda Strain.* I hadn't even heard about it at Scripps; epidemiology was another world to immunologists, after all. But I knew these things had a way of coming back, and when it did, I wanted to be there.

I was getting restless in my professional life and becoming less and less happy in my personal life. I knew it was time to make some serious choices about my future. I just wasn't sure where to start.

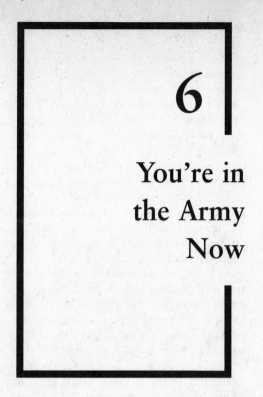

6

You're in the Army Now

AFTER THREE YEARS at Scripps, I felt like I was in limbo. I'd enjoyed the work and learned a tremendous amount, but I really wanted to get back to infectious disease. Scripps was a soft-money place. One of the scientists there was the very distinguished editor of the *Journal of Immunology*. He lost one of his NIH grants and they rolled up half his lab and moved somebody else in practically overnight. Each project was its own profit center and even the big names paid their own rent—a realistic, Darwinian, and capitalistic approach to science.

I wanted to work on T-cell immunity, hemorrhagic fevers, and the relationship of viruses to their natural reservoirs, which just wasn't fundable in that setting. And even if I found money, Scripps didn't have containment labs to do the research.

At this point in my career I also felt I was in danger of being seen as the world's oldest living dilettante: I'd done internal medicine, then some epidemiology/tropical virology, some more internal medicine, and immunology. I was rapidly approaching the point where I couldn't afford to dabble in another specialty and still hope to be taken seriously.

First I tried looking locally and found a position at the VA in San Diego that appealed to me. A virologist there named Mike Oxman had worked with the esteemed and beloved John Enders at the Harvard School of Public Health. Enders was the scientist who adapted polio virus to grow in tissue culture, making both the Salk and Sabin vaccines possible; Mike was starting a program to study viruses clinically in patients at the VA. I had a lot of respect for Mike and the job would have put me in the direction I thought I wanted to go. I'd be in a medical school setting doing research with a small group of first-rate scientists and physicians (the VA was affiliated with the University of California at San Diego). I also thought it might be a good thing for my marriage to stay in the area and not uproot Norma and the kids.

The only downside to the VA job was that it represented another step away from the area I wanted to get back to. People in the scientific community look at you funny if you don't meet the career track; five years of seeing interesting and exotic infectious diseases is nice for cocktail parties, but it's not a line on your CV that counts heavily in the modern medical hierarchy. I remember asking a guy at Parkland for a letter of recommendation and having him write one about two lines long that said, in essence, "I remember Pete very well. He was a good resident and then he went off to Panama and I haven't heard anything about him since." If I was to correct this situation and get on track, now was the time to do it.

I called Karl Johnson at CDC and told him I was looking for a job. Karl immediately invited me over. It was winter in Atlanta, which was a shock after many years living in much more tropical climates. The week of my visit, Atlanta experienced a record freeze, with a big rainstorm that covered everything with a glaze of ice. Pipes everywhere froze and cracked and water was gushing up and freezing on the streets.

Conditions weren't much better at CDC. The building I now work in hadn't been constructed yet and they were still waiting for improved Biosafety Level 4 facilities. Unlike the labs I'd worked in at USAMRIID, the CDC lab had only a hood line for BSL 4 work. I would have loved working for Karl again, but he didn't pull any punches in describing the scene to me: "This is the government. I don't know when this lab is going to be finished. It's overdue and they're behind schedule." Until the new facilities were ready, he explained, I could share a lab with Patricia Webb.

I had a lot of professional respect for Patricia, going back to the days in Panama when she took time out from her life to help me. She taught me a lot about science and research in general. But she was still the boss's wife and she was caught up in her own projects. Realistically, I knew that sharing a lab with her would mean that whenever she needed it, she'd have it, and whenever she didn't, it would be mine. Karl then explained they didn't have a budget for technical support staff.

Karl's moods are cyclic. Perhaps to balance the amazing energy that goes with his periods of scientific brilliance (as when he plotted the course of the VEE epidemic across Mexico and Central America), he also goes through down times. I'd obviously caught him during one of those. Between that, the nasty weather, and the realization that at thirty-seven I'd be a scientist without a lab or a staff, it just didn't feel right.

Around the time I was planning the trip to Atlanta, I got a call from my old buddy Gerry Eddy, who went down to Costa Rica with me during the VEE epidemic. He'd moved up in the military structure and was head of virology at USAMRIID, which was pretty good for a vet (not veteran, but "doggy doc," as we used to humorously call animal doctors). And Gerry seemed anxious to get me in his virology division.

I'd enjoyed working at USAMRIID when I brought back the samples from Cochabamba, but I wasn't sure I was ready to join the army. The labs were state-of-the-art, but there was a certain amount of snobbery in academia that held military-based science in minimum regard. Even though the old offensive biological warfare program had been dismantled nearly ten years earlier, many people still thought of Detrick as "Fort Doom." I wasn't ready to sign up, but I knew and trusted Gerry and I decided to go for a visit.

Going back there reminded me how much I liked the Frederick, Maryland, community. Gerry showed me around the facilities I'd have at my disposal if I went to work there. These were real Level 4 labs, suits and all, fully constructed, with at least some technical support. I also sensed a scientific vacuum, with resources and people plus interesting problems waiting for some enthusiasm and ideas. I listened to the hum of the familiar blowers and was sorely tempted.

Back in San Diego, two critical events took place that helped me make my decision. First, I had a talk with Norma. I knew we were both very unhappy about the way things had been going. For a long time I'd been burying myself in my work (research by day, nights at the clinic) to maintain peace; if we weren't together, we couldn't argue. Research and patient care were perfect, if temporary, forms of escape because they take 100 percent of your attention. This compartmentalization of my life served me well during an epidemic but didn't help in resolving problems in my marriage.

I think Norma, too, had put off the inevitable. So when the time came to figure out where I was going with my career—and where we were going literally and emotionally as a family—it was almost anticlimactic. I outlined the three jobs I was considering and asked her what she thought about them and about moving. Any of the three paid more than I was pulling in at the time, so that wasn't a real consideration.

Norma listened and then said, "I think you should pick any job you want. You should make your choice without any concern about me."

She was insistent: in other words, do whatever you want; the marriage is over. I was worried about the kids. I was very close to all three but I knew it would be out of the question for Norma to consider granting me custody, and in the 1970s it was very difficult for a father to fight for custody unless there was something obviously unfit about the children's mother. I knew I'd miss them terribly, but that would be true whether I moved across town or across the country. On the bright side, at least I didn't have to factor uprooting everybody into my considerations.

The second impetus came from Gerry Eddy. I was on the phone with him during a funding hiatus at Scripps and mentioned how frustrating it was that even though we knew we'd have grant money in a couple of months, everything was on hold. I was working beside Rolf Zinkernagel, who recently won the Nobel Prize in Medicine for his achievements in T-cell immunity. Both of us were very dependent on certain expensive laboratory "supplies." He used inbred mice in his studies and I was using fetal calf serum in my cell cultures. I told Gerry it would be a while before I could order more serum and he made me an offer I couldn't refuse: "Your own Biosafety Level 3 laboratory. You don't have to worry about anybody else in the lab. You'll have a young GI as a technician, as well as a captain who just finished his Ph.D. in biochemistry. And I'll also throw in all the fetal calf serum you can use."

How could I turn it down? I couldn't get the research time or the facilities at the VA that I could at Fort Detrick, and Karl had flat out said I wouldn't have technical support or a lab of my own at CDC. The fetal calf serum was just the icing on the cake . . . so to speak.

Gerry was trying to build a program at Fort Detrick to rival Karl's at CDC. The facilities were obviously outstanding, but he needed to recruit talent.

The only remaining concern was my kids. Before I signed on the dotted line, I sat down with Gerry and the commander at Detrick, Colonel Dick Barquist, and explained that I couldn't afford to do a lot of fieldwork because I wanted to try to get custody of Mayra. I figured I had no chance of getting all three kids, since Antígona and CheChe weren't mine by birth. I was afraid if I was off spanning the globe I wouldn't be able to take care of her and I wanted to be up-front with them about my goals. Besides, I had some ideas about T cells in hemorrhagic fevers. They said that was fine.

The next big decision was whether I'd go in as military or civilian. Gerry said if I insisted he could get me a civilian slot but that would be harder for them and worse for me down the road. They had personnel authorizations for "green suits," but positions were tight for civilians. As a civilian, it would be a lot harder for me to get responsibility and freedom to pursue choice programs (and promotions) later. It seemed to make better sense to join the army, and on July 11, I officially entered the U.S. Army as a major, getting credit for my years of uniformed service in the Public Health Service.

Before I could begin work at USAMRIID I had to go through officer basic training. I reported to Fort Sam Houston in San Antonio, Texas, and was happy to learn of the somewhat relaxed rules for the Medical Corps. For example, if the weather was too hot and the air conditioning was off in the building where we took our classes, we were not required to take physical training (PT). This was fine with us. We figured that we'd already sweated our asses off enough for one day. The barracks were full, so they put us up in a motel down the road, equipped with a swimming pool.

Training wasn't too rigorous either. We learned how to put on our uniforms, where to affix our costume jewelry, how to read a pay slip, and how to stand and march in line. We also had classes on military medicine: half a day on treating nuclear casualties; an hour or so on biological warfare; half a day on gunshot wounds. The highlight for me was two days on map reading, because "an officer is never lost."

After the classroom work we went out to Camp Bullis, where we got to ride around in helicopters and play other neat outdoor games.

In addition to learning how to shoot, we had orienteering sessions, where they'd give us maps and compasses and send us off into the wilderness. Growing up in Texas, I'd shot a .22 and a BB gun, so mastering heavier artillery was no big deal.

When it came time for the night course, though, things fell apart. We worked in groups and I was teamed up with two other guys. One was relaxed and easy to work with, but Bernie was from New York City, and I swear he'd never been off the pavement, let alone in rural Texas at night. It was dark out in the scrub brushlands around San Antonio and Bernie stepped on an armadillo and came completely unwound. The situation rapidly deteriorated as he literally jumped into my arms. I gently persuaded him to let go of my neck and get down, reminding him that, unlike the enlisted recruits, I hadn't gone through rugged PT that would prepare me to carry his body mass and sorry ass back through the brush. Things went considerably more slowly then, as Bernie gingerly picked his way through until we finally made our way back. We were slow but we didn't get lost.

The hardest part of the whole experience was adjusting to the uniform—a far cry from the sandals and tropical wear of my days in Panama. The army says that a uniform is just like wearing a suit and tie to work, but I'd never done that either.

All in all, however, I enjoyed basic training. It was just the distraction I needed to help me through the transition of moving away from Norma and the kids. Even though it was a relief to get away from the parts of my relationship with Norma that wore me down, I missed the kids terribly. I don't know anyone who's been through a separation where children are involved who hasn't been plagued with a persistent sadness and lingering guilt over what things could have been like. When I had time to myself, especially in the evenings, I thought of my Wednesday nights babysitting while Norma worked, and the expressions and pet phrases of my favorite rug rats.

About all I took with me to Frederick were some clothes, books, medical journals, and an orange 1976 Dodge B-200 van. There was no way I would part with the van. Some parents have notches in doorframes indicating their children's heights at different ages; I had pictures of the kids in front of the van at various ages and heights.

Gerry and his wife, Isabel, let me stay with them while I looked for a place of my own. In return, I let them use the van to take their kids on vacation. Unbeknownst to me when I gratefully accepted the invitation to house-sit, the Eddys had a horny little dog named Runto, of perhaps mixed breed but who appeared predominantly dachshund. Runto spent most of their vacation alternately growling at me and trying to copulate with my shin. Although he provided excellent motivation for me to get out, he made the actual search difficult. I remember hopping around the kitchen, trying to hold the phone in one hand and swat the mad beast with the other so I could talk to prospective landlords.

Frederick is a tough rental market, and I was making payments on the home in California and sending Norma money for child support. I lucked out when I found a real estate agent who mentioned a great old farmhouse he might be able to persuade the owner to rent.

After I'd been living there awhile, Ralph and Merle Kuehne, with whom I'd stayed in 1971 when Ralph and I worked on the Cochabamba samples, started introducing me around. (Ralph and Merle later divorced with Merle eventually marrying Karl Johnson and Ralph an Argentine scientist.) They kept telling me about a woman named Susan.

Norma and I had been officially separated for a while and were both emotionally free agents. I felt she was dragging her feet on the actual divorce and division of property, but we'd worked out an arrangement where I'd send her money, she'd send me Mayra over the summers, and we left each other alone.

Ralph and Merle set up the easiest of all blind dates. Susan and I had dinner at their house, friendly turf where we both felt comfortable. Afterward, the two of us went to a movie. I thought that if nothing clicked over dinner, *Annie Hall* would at least save us the awkwardness of forced conversation.

I'd like to say it was just her wit and sense of humor, but I confess Susan caught my attention in other ways too. She had short blond hair, used minimal makeup, and that night wore this nice little yellow-orange dress which was not only very flattering but also coincidentally a warm, friendly, comforting shade that matched the color of my van, which she promptly nicknamed, the "Great Pumpkin." Most important, we had fun together, something I hadn't experienced in a long time.

She was the veteran of a failed marriage herself, so she understood what I was going through. She'd married just out of high school and, as often happens to teenage sweethearts, found that she and her husband grew in different directions, she to a master's degree in library science and he to coaching football. It wasn't a nasty breakup, but she knew how painful it was when two people who'd been in love and planned to spend their lives together couldn't make it work.

After her divorce, she perused the library journals and saw that the Frederick County Public Library was advertising for an assistant director and she applied for and got the job. As we continued to date, Susan adapted well to Mayra's visits, so all the pieces of my new life were starting to fit together quite nicely.

As has been a recurring theme in my life, though, just as things evened out there was an unexpected development. Colonel Barquist had gotten a call from Gerry Eddy, off at a tropical medicine meeting, where he'd just heard a presentation on a disease outbreak in Egypt. Thousands of people were sick, including hundreds of cases of hemorrhagic fever with a high mortality rate. Samples tested at Yale showed it to be Rift Valley fever, a virus active in sub-Saharan Africa but not known to be present anywhere else, including Egypt.

"Look," Barquist said to me when he called me in, "this disease is not supposed to be there. There was some interest in this virus by the United States years ago." I'd been at USAMRIID long enough to know that "interest" was code for "they'd studied it behind the fence." He added, "We've developed a vaccine." Rule One of biological warfare development was that you don't weaponize an agent until you have a vaccine or other treatment for it. Otherwise, if the wind blows the wrong way or something else goes wrong, you're in a whole bunch of trouble.

Barquist explained that the navy had a laboratory in Cairo where they were working with isolates from the epidemic. NAMRU-3 (Naval Medical Research Unit No. 3) needed to protect their people, including American military as well as Egyptian civilians working there. We needed to get enough vaccine over there to immunize them. It could also be an opportunity to field-test the vaccine, perhaps run a trial in communities in Egypt where the disease was being transmitted.

"I know you were to work strictly in the lab when you came here and you're under no pressure to do this, but if you want to take the opportunity to hand-carry the vaccine to Egypt and get some experience with it, we would support that," Barquist said. Barquist was a real commander's commander. As head of the lab, he was a strong, no-bullshit leader. If he said "no pressure," there really was no pressure—except, of course, whatever I developed internally.

I looked at Barquist and wondered what would happen if I blurted out the truth: I'd never heard of Rift Valley fever before, not by name or reputation. I'm not sure I'd ever heard of the Rift Valley. Instead, I took his no-pressure offer as an opportunity to stall.

"Let me think about it for a few minutes," I said, and ran to the library to learn as much as I could about this virus.

The great Rift Valley of Africa runs some 2,000 miles from the Red Sea through the eastern African countries of Ethiopia, Kenya, Uganda, and Tanzania and as far south as Mozambique. This is the site of the Olduvai Gorge and other fossil beds that have informed us so much about the distant origins of mankind. In 1930, after a very heavy rainy season, an epidemic swept one of the European sheep herds there, causing sickness, death, and spontaneous abortion throughout the flock. Veterinarians couldn't get bacteria, however, to grow in the usual petri dishes of media. Yet when they inoculated sheep in their laboratory in nearby Nairobi, they were able to reproduce the disease. They suspected something that was just beginning to be understood as a disease-causing agent, something so small that it would pass through filters that excluded bacteria—a "virus."

Soon enough, shepherds and others near the flocks began to come down with chills and fever; they found themselves unable to get out of bed for days. Then the laboratory workers began to sicken. A man inoculated with the filtered experimental lamb blood also got the disease, and when blood samples taken from him were injected into lambs and sheep, they too came down with the disease. Today, we would question how voluntary and ethical the experiment was, but back then there weren't too many options.

The veterinarians had observed hordes of mosquitoes during the epidemic. Could they be responsible for spreading the disease? They tested that hypothesis by putting netting around some sheep in the affected area and by driving other sheep to altitudes above the mosquito infestation. In both cases, the sheep remained free of infection.

After its initial outbreak Rift Valley fever cropped up in other places in Africa, including South Africa, always causing epidemics in places where there were large herds of European breeds of sheep and cattle. When this disease goes through a herd of livestock, it kills a third to a fourth of the adult animal population and even more of the lambs and calves. All the infected pregnant animals will abort. Even if it misses the human population, then, this disease has a catastrophic effect on the life, economy, and public welfare of a community.

I learned a valuable lesson in public relations when I tried to describe the impact this disease had on rural Africans in the first newspaper interview I ever gave. Weeks after our meeting, Barquist sent me to talk to a reporter from the *Frederick Post.* I jokingly and naively explained, "You have to realize in sub-Saharan Africa a small farmer may have one ox, three wives, and eleven children. He certainly doesn't want to lose one of his wives or children, but it would be a real disaster if he lost his ox and couldn't plow." The guy quoted me with complete accuracy, and I learned quickly to weigh my words and adapt my approach to dealings with the press.

In South Africa, researchers studying the disease had noted you could get infected just by coming into an area and doing a necropsy on a dead bull, which was not an uncommon practice. Someone who lost a prize bull might engage a veterinary surgeon to find out what had happened, and his assistants would get sick. It wasn't clear how these people were infected, whether it was from mosquitoes or contact with the animals or their blood.

There may have been disease in rural areas too, where an outbreak would be less dramatic in the number of cases but perhaps more devastating to a tribal society. But veterinary diagnostics were not very advanced in Africa and sheep death and abortion were fairly common. Unless an outbreak occurred in a large herd where there'd be veterinarians trained to diagnose illness, and unless it infected humans, the virus might pass through unnoticed. Though transmission was by mosquito, no one knew what the natural reservoir was.

In its early appearances in humans, the virus presented only one complication beyond the typical "flu-like" illness: blurry vision caused by a retinal lesion. Patients experienced an inflammation of the blood vessels in the eye. There could be tiny breaks in the vessels, which led to leakage of fluids or even hemorrhaging into the retina. The edema associated with this leakage obscured vision temporarily, but sometimes damage was permanent when blood vessels became blocked so that no blood flowed to the back of the eye in the sensitive nerve layer where it receives light.

As many as 10 percent of people infected with Rift Valley fever experienced these ocular lesions. The more widespread symptoms included fever and malaise. It really knocked people on their fannies, usually incapacitating them for days.

That was our understanding of Rift Valley fever in 1977. But the more I uncovered about this new outbreak, the less it made sense. First there was the location. In addition to Kenya, antibodies for Rift Valley fever virus had been detected in humans as far north in Africa as the Sudan even in the 1930s. There were unconfirmed reports of disease in the Sudan in 1976, just a year before the Egyptian outbreak. But if it came from the Sudan, why did it take so long to hit Egypt? And where had the virus gone between the 1930s and 1976?

On top of that, Rift Valley fever was supposed to cause fever and body aches of a short duration in people, never a fatal hemorrhagic syndrome. In 1977, people started showing up in clinics around Egypt with symptoms including myalgia, malaise, and vision problems. And then there were more serious illnesses: hemorrhagic fever or encephalitis. Understandably, none of the clinicians first thought of Rift; it wasn't even considered as a diagnosis. After all, if you picked up any textbook—including the ones I reviewed at USAMRIID before accepting Barquist's offer—you would have read that Rift Valley fever caused a benign, self-limiting illness in humans with occasional ocular involvement . . . period.

The doctors there thought at first it was Crimean Congo, which causes hemorrhagic fever throughout Africa and would be easier to explain in Egypt. Crimean Congo is a tick-borne disease often carried by birds: ticks cling to birds flying over long distances. Perhaps migrating birds dropped a bunch of ticks in the areas of disease in Egypt.

Another cause of hemorrhagic fever, a frightening new virus called Marburg, after the German city where it was first diagnosed, had a high mortality rate. It had never been seen in Egypt, but it was so new that doctors didn't know enough about it to rule it out until the samples were tested.

It was at this point that researchers at Yale identified Rift Valley fever in the samples from Egypt, a finding so surprising it was presented at the tropical medicine meeting Gerry Eddy attended. It also turned out the South Africans had just linked Rift with hemorrhagic fever during their surveillance for Marburg virus.

It was a unique situation. We didn't know of any other virus that caused such a wide range of illnesses. All viruses have a spectrum of impact in humans, but to have such disparate symptoms was bizarre.

Most people in Egypt had the simple uncomplicated disease. It was debilitating for three or four days with loss of appetite, severe body aches, and fever. As they recovered, perhaps 10 percent developed some degree of retinal involvement. But a small percentage developed severe hemorrhagic fever or encephalitis. They were like totally separate diseases. You might expect more serious cases of hemorrhagic fever to have the retinal vascular lesions and develop encephalitis but that's not what doctors saw. No patient history indicated that people who recovered from one form of the disease ever had the symptoms of the others. This was really strange.

As I digested the information, I balanced the pros and cons of going to Egypt. Probably a lot of sane, rational people would find this a no-brainer: head into a Third World epidemic where people are dying? No, thank you. But I figured we had a vaccine that should protect me (although I'd not yet been vaccinated), so danger from the virus should be minimal. As for my personal life, Mayra wasn't visiting then and this was an assignment that would take days or weeks, not months, so it wouldn't interfere with my pursuit of custody. I'd be gone barely long enough to send a postcard back to Susan.

This was a fascinating disease in a part of the world I'd never been to before. There was an opportunity for a scientific breakthrough if we could use this vaccine to protect people and figure out how the virus got there and why it was causing these disparate symptoms. I told Barquist I'd like to go.

Once the wheels were set in motion I was shocked at how quickly the trip came together. When I agreed to go I didn't even have an official passport, let alone official travel orders. I knew if you needed to go to Kansas, it would have taken weeks to get orders. Here, we were talking about international travel with clearance through the Pentagon. But Barquist was determined to get me out there, and I had my orders, passport, and shots all within two days. He'd personally taken my picture down to the Pentagon to get the passport.

In fact, things moved so quickly that I was completely spoiled upon my arrival in Cairo, where I was immediately transported back to the real-world trials and tribulations of work in an epidemic. I hadn't thought about it, but I was an army research scientist heading to NAMRU-3, a navy lab. Naively, I thought of us all as military. The navy had let me come because they needed vaccine to protect the lab crew from a deadly disease, but they weren't particularly happy to have some army interloper poking around in their territory.

The head of the lab was a physician and navy captain named Ray Watten. He was a smart guy who'd been around a long time and was getting ready to retire. Ray was very hospitable and helpful. I explained the need to vaccinate his people and how the vaccine I brought had prevented lab infections behind the fence at Detrick and that even though we'd had no recent experience with it, I'd gotten it myself three days earlier and still had full use of my arm.

He gave me free rein and I organized what we call an open-label trial. The "open" part in the name comes from the fact that everybody involved in the test knows they're getting a drug. There are no controls or placebos involved. In a lab setting like NAMRU-3, where we were sure everybody would have a lot of exposure to the virus, it made the most sense.

We also contacted the Egyptian Ministry of Health and offered to work with them to vaccinate people at risk in their communities. After several days' delay we finally got to meet with officials there. The ministry itself was just what you might imagine: an ancient building, beautiful marble halls, high ceilings, but with plain, often shabby furniture. Cooled only with fans, the building was like the inside of a hair dryer. After several meetings it became clear the Egyptians weren't interested in our vaccine or in granting a permit to let us study the disease in the field.

There were several problem areas to deal with. For one thing, this vaccine was experimental. It was approved for investigational use by the Food and Drug Administration but was not yet licensed. We need FDA approval before we're able to dispense drugs or vaccines even for experimental or investigational use. Some may quibble with FDA policies, and it's true that good drugs don't get to the market as fast as they might. But by the same token, ones that prove dangerous are intercepted before they can do a lot of harm. We have only to look at the frightful legacy of thalidomide and compare the situation in the United States with that in Europe, where thousands will suffer the effects of the lack of sufficient testing for their entire lives. I remember as a resident reading about new drugs in the British medical journal *The Lancet* and wishing they were available to us, only to read later of their withdrawal from the European market because of previously undiscovered side effects. (In reality, there are no such things as "side effects." Potent pharmaceuticals have certain effects on the body. Those that we want, we classify as effects. Those that we don't want, we classify as side effects.)

Even outside this country FDA approval carries a lot of weight. Wherever you go in the world, if you want to use a drug or vaccine people have never heard of before, they will ask, "Is this approved by your FDA? Is it licensed in your country?" I've found in general that if you say the FDA of the United States has licensed it, you'll be able to use it. If it's not, you'd better be able to explain why not. Our vaccine was still in the experimental phase because extensive testing hadn't yet occurred. That was probably one reason Egyptian authorities were very gracious and friendly but politely declined our offer.

At the time the United States and Egypt were only just beginning to stabilize their friendly relationship, which President Jimmy Carter would soon broker into a formal peace. When some guy from the U.S. Army showed up with a drug only approved for experimental use, it wasn't an easy sell.

Finally, the Egyptians had a lot at stake if they officially recognized the seriousness of the outbreak. If word got out, it wasn't going to be good for tourism. Furthermore, this was around the end of the monthlong Muslim fast of Ramadan, when sheep are traditionally ritually sacrificed, and Egypt exports a lot of sheep to Saudi Arabia. No one came out and said, "If too much negative press gets out about this disease, we'll be ruined. Go home!" but that was certainly an underlying concern. Authorities wanted to stop disease in their country but every step they took threatened their economic situation. Officially, then, I was only there for consultation. Within a few months, however, the situation resolved itself. So many animals died that Egypt effectively had nothing left to export.

With no Egyptian government cooperation or authorization to study the disease, I focused all my energies on getting the folks at NAMRU-3 vaccinated. First we needed a consent form in English and Arabic so American and Egyptian populations could understand it. I'd never put together a consent form before but I wrote one up and had it translated. Then it had to be blessed by NAMRU's human-use committee.

Finally, we went before our target population to explain the vaccine to them. This was easy to do with navy personnel: "This is a vaccine developed by the Department of Defense. We've used it in several hundred people and we think you ought to have it."

Their response: "Yes, sir. Thank you, sir." It was harder to explain to the Egyptian workers, for whom everything had to be translated not just linguistically but culturally. That took the better part of a day, but we finally got everybody ready to go.

The first step, as in Costa Rica, was to get blood samples from everyone beforehand to make sure they didn't already have Rift Valley fever (a current infection or antibodies from a previous exposure). Then we vaccinated them and took blood samples at intervals to check their immune response. One thing the army didn't know was how soon the vaccine would provide protection. How fast would people form antibodies?

We also wanted to find out how effective the vaccine was. Vaccines don't just work or not work; they protect a little bit or a lot. At 50 percent or less effectiveness, for example, you're not talking about a viable vaccine. Some vaccines are 99 percent effective. The influenza vaccine is estimated to be between 70 and 90 percent effective, certainly well worth having. We hoped the tests and sera in Egypt would help us develop a good estimate for the Rift vaccine.

We also had to get a handle on the disease itself: who got sick, how many people got sick, and how sick did they get? To do this, we had to go out in the field and gather information from places affected by this outbreak.

The most intense disease transmission was originally in Upper Egypt, near Aswan and Asyut. Since the Nile flows from south to north, the southern part of the country has always been referred to as Upper and the northern part, near Cairo and the Delta, as Lower. The disease then skipped over some areas. There was a lot of intense spread through the Delta region as well, particularly around the town of Zagazig, where Jim Meegan made the first isolate of the virus causing this epidemic.

Since Zagazig seemed the epicenter of the epidemic, that looked like a good place to get blood samples, talk to people, and get an idea of the magnitude of the disease. As we traveled through Egypt, I saw diseases like pellagra and beriberi, which I'd never seen in the United States except in the most absolute, die-hard alcoholics. The Latin Americans were not rich by any means, but I didn't see the extent of illness caused by vitamin deficiencies there that I did in the poorest areas of Egypt.

Since there were no official figures provided, we used our own methods to calculate the impact of the outbreak. One way to do this is a method called excess mortality. That is, we looked at the number of people who died in a particular town in a similar period in the preceding year and compared it with the number of people who died this year. It's not as exact as we'd like, but it can provide a good estimate. After counting the dead in the Bilbeis town registry, it seemed that about 1 percent of the people who lived there died in the year of the epidemic. Less than 0.5 percent of the people who lived there died the previous year, so mortality was up at least 0.5 percent.

We also got an estimate from an Egyptian general who ran a big military training facility near Zagazig. He said he'd done some studies and also found a mortality rate of about 0.5 percent, although he wouldn't describe the studies or give us any numbers because that information was classified. Even with the peace efforts being put forward by the United States, the Egyptian military wasn't ready to provide the number of troops in an area near the Sinai.

In Bilbeis, the people we interviewed gave us some interesting information that we hoped might help clear up part of the mystery surrounding the epidemic. We had no explanation for the pattern of transmission. While we were talking, a large aircraft passed overhead, readying to land nearby. It turned out the airfield was just over the horizon, and there was an army base near Bilbeis—which may have been the one our friend the general told us about. People remembered hearing that the base had been sending people to sub-Saharan Africa.

Soldiers will bring back just about anything they can get their hands on as a souvenir. We mentioned this to the townspeople and they remembered some of the soldiers had African monkeys with them. We didn't know what the reservoir of Rift Valley fever virus was, so perhaps this was a lead. Even if the monkeys proved a dead end, the soldiers might have brought back infected sheep or even an infected mosquito.

The soldiers themselves were candidates for transporting the disease. Once the virus was in their blood, they could have been bitten by a local mosquito which could pass the disease on to others. If the soldiers were infected, they wouldn't have spread it to locals directly, because it wasn't transmissible person-to-person like the flu.

The one thing we suspected for sure was that there was plenty of troop movement in the region, even if no one would confirm it. All Egyptian soil remotely near the Sinai was full of Egyptian military ready for another skirmish with the Israelis. The whole area was said to be a big military zone, and it was rumored that there was an Egyptian special weapons manufacturing facility nearby.

One theory about the outbreak was that the Egyptians had weaponized Rift Valley fever, or the Russians weaponized it and then left some behind as a stabilized powder or a frozen liquid. Viruses are chosen as potential biological weapons because they're infectious by aerosols that can be carried on the wind. Perhaps the Egyptians had some Rift that somehow got out of containment. Maybe a lab worker was infected and had been bitten by a mosquito that could transmit the virus to others. Egyptians on the political fringe were spreading rumors that the virus had been sent by Qaddafi from Libya as punishment for opening a dialogue with the United States. Others thought it came from Israel.

There were a lot of scary possibilities which couldn't be definitively proved or disproved. The disease spread so efficiently (it could very quickly blow a herd of sheep apart and infect all the people nearby) that unless you caught some guy up on a hill with a blower and gas mask generating aerosols from a can, you couldn't be sure if it came from man or nature.

Less conspiratorial theories centered on ecological changes resulting from construction of the Aswan High Dam and the resulting formation of Lake Nasser. One scientist argued that small arthropods like mosquitoes and sand flies could get sucked up into air currents—particularly in areas around what's called the intertropical convergence zone—and carried for long distances. In other words, infected mosquitoes from the Sudan were swept by air currents into Egypt and came down and bit the sheep. To me, that seemed a little far-fetched.

I thought it more likely that the spread centered on the movement of domestic animals. There was a railroad line from Khartoum in nearby Sudan that connected with a ferry at Wadi Halfa, just on the Sudanese side of the border, that took you right up to Lake Nasser and a ferryboat ride to Aswan. There were various quarantine stations at different points on the trip, but it's likely that the quarantine was bypassed with some frequency. There were also roads running parallel to the Nile throughout the country, so it was easy to imagine how an infected animal, person, or mosquito could have gotten from one area of the country to another. A fellow scientist studying foot-and-mouth disease once showed me a photo she took of a sheep riding in the sidecar of some guy's motorcycle, heading south on the road out of Aswan.

Timing would be tight, because in domestic animals you've only got two or three days' incubation and three or four of viremia. An infected animal would have to have been transported quickly to one of the affected areas for a local mosquito to get the virus, but it would be possible and seemed the most likely explanation.

I stayed in Egypt for three weeks in all, overseeing vaccination and testing at NAMRU and making unofficial forays into the field. I traveled either on my own or with a CDC epidemiologist named Dave Morens, who was facing similar obstacles studying disease in the country. We'd wander around a bit and then stop by the embassy commissary and buy some brandy. If we couldn't study disease in the field, we were at least going to have some fine philosophical (if somewhat drunken) debates, usually centered on which was more important, epidemiology or the lab.

I found the average Egyptian on the street tremendously helpful. If I was lost I could ask anybody for directions. If they didn't speak English but understood where I wanted to go, they would literally take me by the hand and walk me to my destination—often well out of their way. Despite its poverty, Cairo was probably the safest big city I'd ever been in. You could be anywhere in the city at any hour— man or woman, local or foreigner—and you felt safe. If a modestly dressed, unescorted woman was being bothered by some man, other men would naturally rise to her defense. I found Egyptians to be among the kindest, friendliest, and most giving people I'd ever met.

At the same time, working with them could be incredibly frustrating. If you planned to meet somebody at 5:30 they would definitely be there at 5:30—if their car didn't break down or there weren't other commitments they'd forgotten or something new hadn't come up that demanded their attention.

After about three weeks, everyone at NAMRU-3 had been vaccinated, follow-up sera had been collected, and it was time to return to the States and start work on the virus.

The motto of the Medical Research and Development Command at USAMRIID was "Research for the Soldier." It was our job to protect the armed forces and conduct research to protect them in the future. Navy personnel at NAMRU-3 had an immediate need for protection against this virus, so I brought them vaccine. This served long-term goals as well since it gave us more experience with the vaccine out in the field.

The Rift vaccine is a killed vaccine, like the Salk polio vaccine. Unlike when using a live-attenuated vaccine—injecting a diluted amount of attenuated virus that will grow enough to stimulate an antibody response—you need a bigger dose of a killed vaccine to get an immune response. Thus the amount of Rift vaccine was limited. We couldn't give it away haphazardly.

We could provide vaccine for people or groups at highest risk. If we got a request from a vet going to work in sub-Saharan Africa, for example, a person who would be at risk of high exposure, we'd provide vaccine. We also sent some to CDC and some of the Egyptian laboratories. After an Egyptian lab worker was infected and developed blindness, there was a different attitude toward the vaccine. Realistically, aside from Egyptian scientists and lab workers who were going to be up to their armpits in virus research to try to control the disease in their country, it was cost-prohibitive for us—and certainly for their government—to vaccinate everyone. In the 1970s, we

costed it out at five dollars a dose, or fifteen dollars per immunization—hundreds of millions of dollars to protect the entire nation.
That's enough to get your attention in the United States and certainly
in Egypt.

There were two funding streams at USAMRIID related to viruses.
One was called infectious disease (although all viruses studied caused
infectious disease) and involved natural threats. If we were fighting
the Spanish-American War today, for example, we'd need to protect
troops from yellow fever. The other funding stream was biological
warfare and defense, which meant protection against planned attacks
using weapons which might employ natural or altered disease agents.

As often happens when two related divisions compete for funds in
the same umbrella organization, minor wars erupt over turf. My trip
to Egypt highlighted one of the problems in differentiating between
the two. Technically, we brought vaccine over to protect U.S. military
personnel, which, because the vaccine was developed for biological
warfare defense, meant funding should come from the biological
warfare side of the house. But this was a disease that occurred in
nature and we had no belief that the outbreak was anything other
than a natural epidemic.

There was internal dissent about the role of researchers working
on biological warfare defense. Some felt we should spend all our time
developing vaccines and antiviral drugs and never leave the lab, let
alone the country. To them, Egypt was just a distraction. If it was a
natural epidemic, it should be done by infectious disease people with
their money, not by biological warfare defense funds.

But infectious disease wasn't doing it and we needed test beds for
vaccines, diagnostics, and drugs. We had personnel trained in all the
disciplines involved, from defense to microbiology to public health. It
was more important to get people up to speed in recognizing this
disease than to waste time fighting over who should deal with it—or
worse, doing nothing.

From a military standpoint, if our troops were attacked by BW,
we'd need to understand the spectrum of the disease so we could tell
a commander what to expect if they were exposed to Rift Valley
fever: whether they would live or die; how long they would be sick;
what the attack rate would be. There were a lot of functions served
by the study of this disease in nature.

Also, the hemorrhagic fever and encephalitis symptoms we'd seen
with this Rift outbreak made us wonder if we were dealing with a
different strain, one we'd certainly need to understand before we

could develop effective countermeasures to prevent and/or treat it. I was anxious to get back and isolate a good source of virus that we could use for vaccine testing and study in the future.

A couple of issues had to be resolved before we could begin research back at the lab. The immediate concern was getting samples into the country. Even before I left for Egypt, Gerry Eddy's job was to convince the U.S. Department of Agriculture to allow us to start up a Rift program. Officials there shit the proverbial brick. Rift Valley fever was number six on the list of viruses the USDA most wanted kept out of the country. Gerry was finally able to convince them that the chance of this virus getting loose outside our facilities was virtually nil.

The next question was more philosophical: how best to study the virus, how to determine the ways it was different and whether or not our vaccine still protected against it. Back in the days when Fort Detrick's offensive program was active, they would induce the disease in human volunteers when they knew the agent caused a mild, self-limiting disease or was susceptible to treatment by antibiotics. For example, they determined the aerosol-infectious dose for agents like tularemia or Q fever by placing volunteers in huge tanks where an aerosol could be released. The scientists charted volunteers' respiratory rates and calculated how much air they were breathing and how much agent would be deposited in their lungs. Tracking the dose of aerosol, they could actually quantify the minimal infective dose. Volunteers who weren't protected and got the disease were quickly given antibiotics to treat the infection.

But here we were dealing with a virus not treatable by antibiotics (as most bacterial infections are), which caused a disease we knew could be fatal. This could be a new killer strain of Rift and we couldn't take it for granted that the old vaccine would be effective against it. We sure as hell weren't going to induce Rift Valley fever in people to find out. I'd be able to test it against the antibodies people at NAMRU developed after being vaccinated to judge the limits of protection, but initially we couldn't take any chances.

In addition to the vaccine, we wanted rapid diagnostics so we could see what we were dealing with right away. These would have to be sensitive but also real-world. They had to be adaptable wherever an epidemic or attack occurred. We eventually settled on developing an ELISA test. Other tests might have been more sensitive, but they were not practical for use in the field. In one competing method, they could detect amounts as small as ten virus particles, but the process required distilled water and perfect conditions to get an accurate reading; it failed miserably in the complex mixture of a patient's blood. We wanted eventually to develop an immunological dipstick test, similar to a home pregnancy test. Since antibodies are the key element in the body's recognition and defense system, they are a natural base for these tests.

Antibodies do a job for the body—they bind to things that are foreign and help get rid of them: by neutralizing a virus, helping macrophages eat them, or activating a series of proteins in the blood serum called complement. The foreign "invader" they bind to is called the antigen. We can measure antibodies by taking advantage of the binding property. There are three common ways to do this.

One way looks at the functional changes the antibody induces in viruses—virus neutralization. In this test we mix serum with a known amount of virus, incubate so the antibodies have a chance to attach to the virus in the solution, and then measure the amount of infectious virus that remains. If most of the virus is no longer present in our assay, then we say that we have a serum with neutralizing antibodies. There is a very good (but not perfect) correlation between the presence of neutralizing antibodies and protection from virus infection.

All antibodies in serum are not necessarily able to neutralize virus, and we often use other tests that measure directly the binding of antibodies and do not require any functional properties of the antibody.

The simplest binding test is the fluorescent antibody test (FA), in which the reaction is revealed because the antibody has been "tagged" with a dye that fluoresces when black or ultraviolet light is shined on it. In this test we use virus-infected cells that have been fixed on a microscope slide. We can't see viruses with an ordinary microscope, of course, but we can see aggregates of viral antigen in the cells if they are revealed by a fluorescent antibody test. When we treat the cells with the unknown serum, specific antibodies will bind to the viral antigens and remain stuck as we wash away all the other antibodies. To reveal the bound antibodies we take advantage of the differences between species of animals by injecting human antibodies pooled from many people into a goat or rabbit, which will make an immune response against the parts of the human antibodies that differ from the animal's own. (It's tricky to keep this straight at first because the human antibodies have become the antigen for the rabbit's immune system.) These rabbit antihuman antibodies, tagged with the fluorescent dye, are reacted with the slide to build a "sandwich" of viral antigen + human antivirus antibody + tagged rabbit antihuman antibody. After washing away the tagged antibody that didn't bind to the human antibody, we can examine the cells under a fluorescent microscope and see the glowing viral antigens against a dark background.

The last test is the enzyme-linked immunosorbent assay (ELISA). It has largely supplanted the fluorescent antibody test today because it is more objective, faster, cheaper, and can be semiautomated readily. However, the principle is the same. We attach the viral antigen in the bottom of one of ninety-six wells on a small plastic plate and then allow the unknown serum to bind to the antigen. After washing, we complete the sandwich with a rabbit antihuman antibody, but it is tagged not with a fluorescent dye but with an enzyme. The job of this enzyme is to reveal the binding of the antibody by reacting with a colorless liquid we add last to develop color, which is what we measure.

In addition to a diagnostic test for Rift, we needed a realistic model on which we could study therapeutic interventions. Before jumping to human testing, we start out with smaller animals to define what treatments might be realistic and helpful. Once we've identified good candidates, we can then test them in animals, often monkeys, that best mimic the reactions to the virus experienced in humans.

To get started, we had to immunize some people at USAMRIID. I'd been vaccinated before I left, had no negative side effects from the vaccine, and hadn't caught the virus in Egypt. We couldn't guarantee protection, but it was definitely better than nothing. Initially, only I worked with the Egyptian virus, until I'd run enough tests to feel confident that others would be safe once vaccinated.

I was almost obsessive-compulsive about documenting all the protocols. I wanted to keep track of everything we did, from initial virus isolation through any and all experiments we ran. At the end of our research, if we had a viable vaccine that we wanted FDA approval for, I wanted every step recorded in a lab notebook so we could show them exactly what we did.

First, I tested the antibodies from the sera taken from people vaccinated in Egypt to see if they were effective against the new virus. In these neutralization tests, the new Egyptian strain was neutralized to the same extent as the familiar viruses from sub-Saharan Africa, suggesting that the virus from this outbreak was antigenically much like the virus in earlier outbreaks. The vaccine should protect against it.

Scientists don't like to draw a conclusion from a single set of data, so I'd also started some tests with mice. I vaccinated half of the mice and then split the vaccinated and unvaccinated into two groups: one challenged with the old virus, the other with the new. The protection in mice was similar, which reinforced my confidence in the vaccine. We were ready to bring more vaccinated people into the labs.

One goal was to compare the differences between this virus and the earlier Rift virus to try to understand and explain the different symptoms in humans.

We worked up the Rift experiments in an augmented Biosafety Level 3, essentially the same as Level 4 except that we wore hospital scrubs, mask, gloves, and cap instead of space suits. To get in the suite you first entered what was called the "cold side" change room off the main corridor. Initially we used keys to get in but later switched to magnetic cards. There were separate cold side changing rooms for men and women, and this was where we took off our street clothes and got our scrubs. From here, we entered a "hot side" change room, which was really just a room with a toilet and bench to stow our lab shoes. When we departed, we left our shoes behind, put our scrubs into a bag to be autoclaved later, and exited through a shower where we washed down thoroughly. Virologists at RIID tended to be very clean people, albeit with chapped skin in the winter.

Eating, drinking, and smoking were not permitted in the lab, but there was one room set off to one side that served as the office/coffee-break room. All rooms were tied to the blower system and independently ventilated negative to the hall. The floors and walls were sealed, the sewage was cooked, and the air was HEPA-filtered before it left the building. It would have been difficult for a virus to get out of there, or get from one room to another. In terms of environmental protection, it provided the same containment as Level 4, but it offered less protection for the workers, since the agent we were working with was protected against by vaccination.

I inoculated mice with Rift Valley fever virus to get the first set of experiments going. Walt Cannon loaded the syringes with various dilutions of virus and I did mouse retrieval and injection. Mice are actually easy to work with. If you pick them up by the tail and place them on a wire-top cage, they'll pull away from you, which puts them in the ideal position to be grabbed by the scruff of the neck.

You have to get enough scruff so they can't turn around and bite you, but then you can pick them up, wrap their tail under your little finger, and you're ready to go. They can't bite you and you can inject them easily.

Next, we did the hamsters. The hamsters, again, were not garden variety pet hamsters. These were tough critters, born and raised in captivity with a bunch of other hamsters where the rule is survival of the nastiest. They are not accustomed to being handled by people and don't adapt to it at all. I'd done enough work with hamsters at MARU so that the hissing and baring of teeth didn't intimidate me.

The last set of animals, however, had me spooked. I'd saved the rats for last because I hadn't worked much with them. They were also the biggest, with the biggest teeth. They're simply the animals most likely to make the hair stand up on the back of your neck.

"Okay, this is going to be a two-man job," I told Walt. "Are you gonna hold them and I'll inject them or am I gonna hold them and you inject them?"

"Either way, Major," Walt replied. He was enlisted and I was the officer, so he deferred to me. A good soldier, Walt had volunteered to fill in when I suddenly lost my regular lab technician to pregnancy (pregnant women were not allowed in the Level 3 or 4 labs).

I said, "Well, how do you hold rats?"—thinking I'd find out from him how to do it.

"I don't know," he responded. "I'm a medical laboratory technician. I deal with humans. I've never injected an animal before."

I tried picking them up the way you pick up a mouse, but rats are resourceful. They'll turn around and climb up their own tail, so you don't have much time. When you grab them by the scruff, they're strong enough to double their body over and reach you for a juicy bite. After a few experimental, aborted attempts, we ended up with my grabbing them by the scruff and tail, while Walt injected them.

We had maybe fifty to seventy of each of these animals, so inoculation was an all-day affair. The whole situation illustrated to me the compartmentalization of resources in the government. There were plenty of people around who could have done it more easily than we, but it was my project and my job to do it. The timing was off; it wasn't a big year for my department from a resource standpoint. This was just an introduction to a phenomenon I'd find over and over again in a bureaucracy. Sometimes a relatively small project can be a tremendous struggle to get through, not because the resources don't exist and not because people don't recognize the importance of the work, but because somebody else's department has the goods—or people, or funding—that year. And once you've gained ground in the race for budget, you don't want to give any of it up.

It took a lot longer to get things going than I'd anticipated. When we started developing good data, though, some interesting points emerged. We noticed, for example, that Rift Valley fever virus from Egypt (called Zagazig 501) had very similar pathogenicity for mice and hamsters as the sub-Saharan strain (known as the Entebbe virus). Results were similar in rabbits and gerbils, which seemed to indicate the two viruses weren't very different from each other.

But when we got to the rats there was an enormous difference. Egyptian virus killed rats at very low doses while the Entebbe strain needed a million times more virus to kill them. This is a pointed illustration of the specificity of the virus/host relationship. In the other rodents, there was no noticeable difference. But this particular type of rat (called a Wistar-Furth rat) grew the virus very easily and was highly sensitive to it. Virus reached tremendously high titers in their blood and they'd die in two or three days. The vaccine protected the rats, as it did the other animals, but without it they were dead. So was the Egyptian strain simply more virulent?

We still needed an animal to provide a good model for the human disease. Mice, rats, hamsters, and rabbits were a good starting point, but they weren't realistic models. The rats probably came the closest in that they experienced the full range of disease exhibited by the people in Egypt. Some rat species were more resistant or sensitive than others, but we were able to detect the transient virus circulation terminated by a rapid antibody response (which would correspond to the benign form of the human disease) in some; in others we found the same benign course followed by encephalitis; and yet others came down with a raging, progressive viremia that had some similarity to the hemorrhagic fevers that occurred in humans.

The rat studies showed that reaction to the virus was animal-specific, which underscores the importance of using a variety of animals in initial tests. If we used only mice and hamsters, we wouldn't have observed the dramatic difference in reaction and would have been less inclined to believe the viruses were that different.

With any work that involves animals, you have to be sure the knowledge you hope to gain is worth it, in terms of both the animals' life and health and the lab workers' safety. These days, every protocol is reviewed by animal-use committees, some of which are made up of people who may not have the scientific competence to fully understand a project's goals and procedures. This can be frustrating, especially if you're in the middle of an outbreak, desperate to test treatments on animals before you start injecting drugs in sick and dying people.

In addition to the safety hazards, it's not easy to work with animals. You have to be completely focused on the task at hand and yet you have to remove yourself emotionally from the animal. I think again of my grandfather on the farm. He appreciated his animals and kept them as comfortable as possible, always treating them with respect. But when the time came for one to be sacrificed for food, to him that was the natural order of things.

Often, too, the presence of animals poses a direct threat to humans, as with the rodent reservoirs of hantaviruses and Bolivian hemorrhagic fever. Some of the animals I've worked with have led to the use of vaccines that have prevented and will continue to prevent death and suffering in people around the world. In the past five or ten years animal models have given us real insight into the toxicology of drugs and different antibiotics and told us which antibiotics are effective in certain situations. Many people are walking around today because of porcine heart valves. If some human diseases common in this country had been modeled more extensively in nonhuman primates, it's possible we would have more advances and better ways to deal with these viral infections than we do today.

Having said that, I really don't like working with monkeys, and it's not just because they look too much like me in the mirror in the morning. You don't go into a lab full of monkeys and start making friends with them. When there's a monkey making a threat display at you, you can intellectually understand that he's trying to keep you at a distance, that he wants his own space. But at the same time, you know you have to work with him and that doesn't make it any more pleasant.

They're also dangerous. They're tough, smart, aggressive, and can bite the hell out of you. Rhesus and cynomolgus macaques carry a chronic infection called simian herpes B. If they're shedding that virus and they bite you, you will probably get encephalitis and die. Just another reason to be sure what you're doing is worth it.

Nonetheless, our next level of tests involved working with rhesus monkeys. In the interest of safety, I made it a policy that, unless completely unavoidable, nobody worked with monkeys alone. I also insisted that we use what are called squeeze cages: cages whose backs you can move forward by pulling a couple of poles in the front. Using these, you'd nudge the monkey up against the front of the cage and then he'd be close enough to receive an injection while his freedom of movement (and biting) was inhibited by the temporary restriction of the shortened cage. We'd typically knock the monkey out with an injection of ketamine. Then it was safe to take him out of the cage, get a blood sample and baseline vitals, inoculate him, and return him to the cage before his head cleared completely.

I think it's important to say a few words here on the subject of animal research, because it's one of the most sensitive and controversial issues in all of medical science.

Many of the procedures we do are the same ones we then do with human volunteers. We do them with monkeys first because we place the lives of humans first, but eventually the same treatment is extended to humans.

Experiments in which you're injecting a hazardous virus that you expect will make the monkey sick and probably even kill it obviously carry much more negative impact for the monkeys, but are vitally important. The less we know about a given disease, and the more serious it is, the more valuable and necessary animals are to our quest for knowledge.

I have heard arguments about using computer models instead of animal models. I'd love to use computers, but no one has developed the software that replaces an animal model. Maybe one day they will. The endless array of variables in a living organism—human, monkey, rat, mosquito, whatever—have so far made it impossible to generate an accurate computer model.

As for bypassing animal experiments and testing directly in humans, when I became a physician I took an oath first to do no harm. Just as I have to be convinced the use of an animal is warranted, I need a base level of understanding of possible repercussions of my actions before I do anything experimental with a living human being.

I don't have a problem doing experiments with animals in a well-regulated system, where the burden is on me and my colleagues to justify their use before committees that oversee my work. But the same burden of proof needs to be applied to alternative ideas. If you believe an experiment is unnecessary or can be modeled successfully on a computer, show me exactly how. I'd be delighted to use it.

But we didn't have that alternative with Rift, so we began working with rhesus monkeys. They provided a good model that gave us insight into what happens to people when they get exposed to the virus. We developed this understanding by monitoring the immune response of the monkeys at different stages of infection and observing how this corresponded with the symptoms they presented.

The immune systems of both monkeys and humans are comprised of a complex and impressive array of defenses. The systems also manufacture antibodies, which react specifically to a particular virus or other antigen, in an attempt to neutralize the threat before it has entered a cell. Then there are the agents which fight for us if the invader has penetrated behind our front lines—that is, if the enemy has already taken over the cellular reproductive mechanism for its own purposes.

T cells are so called because they mature in the thymus. If the antigen is already in the cell, they destroy it, even at the expense of the cell itself. This "scorched-earth policy" is why some T cells are referred to as killer cells. They also secrete very potent molecules called cytokines with many different defensive and also physiological actions. The actions of cytokines such as interleukin 1 help explain why we often feel tired and weak and fatigued while "fighting off" a virus.

These T cells also have responsibility for policing changes in the body, such as may occur when new antigens appear on cancer cells.

Antibody defenses are mounted by the B cells, named for the bursa of Fabricius, part of the body in chickens where they were first identified. They hang out in the lymph nodes and spleen, waiting for something to come to them to indicate there's a problem. Local activity in these sites is probably the reason the spleen is a good organ to get hemorrhagic fever virus samples from and, in the case of lymph nodes, why you get swollen glands during an infection.

A fetus has only a set of B cells, each of which recognizes specific antigens through the expression of its particular antibody on its surface. You develop more B cells, including ones that will be specific for additional antigens, in a complicated process whereby these cells make new cells that bear different mutations of their antibodies. A very talented immunologist named Mel Cohen used to refer to this process as the Generator of Diversity—GOD. GOD helps you develop new B cell specifiers so you're prepared to deal with more foreign substances. Throughout your life, when a B cell is triggered by the presence of a new antigen, it divides as it changes into a state where it secretes the antibodies that actually fight the virus. This is all a gross oversimplification of an amazing process, but it gets across the general idea.

As you're exposed to more and more viruses and other foreign agents, and develop more and more different antibodies, the composition of your blood is altered slightly.

A newborn, germ-free baby, for example, has very low immunoglobulin or antibody molecule levels. If you are raised in a middle-class family in America, by the time you're an adult you'll have something like ten milligrams of immunoglobulin per milliliter of blood. But if you've been raised in an African village where you've been repeatedly exposed to a variety of parasites, viruses, and the like, you may have twenty or thirty milligrams of immunoglobulin in that same milliliter of blood. You also have tucked away memory B cells that can be called upon to give a new batch of antibodies as needed.

One of the mysteries we're still trying to unravel is the question of whether you need "antigen persistence" to drive B-cell memory, which strikes at the core of long-term vaccine efficacy. Some people say once you've made the memory B cell, then no problem, you're set for life. Others argue that your immune response will last only for a while and that you'll need an antigen boost now and again to keep the memory alive. This is why people are advised to keep their tetanus shots current—every eleven years—because this lethal disease is still a threat in the United States and we don't want to take chances on memory loss.

There are also nonspecific ways of fighting off foreign invaders such as macrophages, which are essentially "commando" cells that chew up and destroy anything that doesn't look like a normal body component. The macrophages are quite talented and have a variety of cooperative activities with T cells, B cells, and antibodies.

Another player in the body's immune system is interferon, a small protein molecule that is often made in response to viral infections. The presence of viral nucleic acid induces secretion of interferon, which can be produced in almost any cell in the body that's suitably stimulated. Interferon in a cell creates what we call an antiviral state. The cell will be harder to infect with virus and if it is infected, it will produce less virus. Unfortunately, interferon therapy for virus infections hasn't proved terribly helpful because by the time people came in for the treatment they'd usually already produced as much interferon as they could use. Interferon has proved useful in treatment of some chronic viral infections such as our old friend hepatitis B or the newly discovered hepatitis C. Interferon also helps the T cells in their work by making infected cells easier for them to recognize.

As we studied the monkeys, an interesting pattern arose involving their response to Rift. It gave us what looked like our first clue as to why some people got very sick in Egypt and others had only the milder form of Rift Valley fever, even though all were exposed to the same virus. The sequence of events in monkeys seemed to be that after they developed an infection, they made an interferon response to the virus. The critical point was how soon after infection the interferon kicked in. If they made an interferon response within the first twelve hours, they had a mild disease. If it took up to twenty-four hours for a detectable interferon response, they'd get sicker. And if it took thirty-six hours or longer for them to make an interferon response, they'd have severe hemorrhagic fever and they typically died.

We found no correlation between the strength of the response and the disease outcome. Some monkeys actually had a very poor interferon response, but it started quickly. With this disease, timing was apparently everything. On the other hand, in rats it was the strength of the antiviral response from interferon, rather than the speed, that was critical. Again, this shows the importance of not relying on a single species for implications to human disease.

The critical event in monkey disease, and therefore probably in humans too, is that the interferon system had to stay a step ahead of the virus. If the interferon wasn't out there protecting the endothelium (the lining of the blood vessels), the patient was in trouble. Healthy endothelium has special properties that keep the blood from clotting while circulating in the body. When you cut yourself, or spill blood on the floor, contact with the foreign surface initiates clotting.

Rift Valley fever virus is very destructive to cells. If you introduce it in cell culture, depending on how much you put in, it can strip all the cells off the surface in a period of twelve to forty-eight hours. We call this damage cytopathic effect (CPE). Some viruses, like hantaviruses or such arenaviruses as Machupo, may cause little or no CPE to cells. But with Rift, the CPE is so dramatic that you can easily see disruption and death of the cells under a microscope.

We hypothesized that Rift had the same cytopathic effect on the endothelial cells of an infected person (or monkey) that it did in cell culture if an interferon response wasn't generated quickly enough to stop the virus before it got there. Once it infects the endothelium, it destroys the cells, causing microvascular clotting all over the body. This happens primarily in small blood vessels but can go all the way up to the arterial level. And once the cycle begins, it continues.

The microcoagulation continued in monkeys even after their blood had cleared of virus. There had been so much damage to internal organs from intravascular coagulation that in some cases they continued getting worse for three to five days after the "infection" was over.

About half of all people who got hemorrhagic fever from Rift died, a terrifying mortality rate. If we could confirm this intravascular coagulation, maybe we could make a difference in the prognosis.

This opened up a whole new area of research, in which a class of drugs known as immunomodulators were developed to stimulate the immune system, often through interferon production. Frankly, I didn't think they were worth a damn, but the army was very enthusiastic about the research. And since I wanted to know more about the use of newly available biotech interferons, I figured I might as well dovetail our interests.

We continued to back up the monkey experiments with extensive studies on smaller animals, including rats. While I'd gotten fairly adept at handling the rats, I still didn't like it. Thus, when another researcher offered to show me his techniques, I was happy to observe.

"I'll show you how to handle these rats," he said, donning a steel-mesh glove.

Just as he picked up the rat with the gloved hand, someone walked in the room and called out, "Telephone, Bob." When he looked up, the rat bent over itself and chomped his other hand to the bone. This was an uninfected rat, and we were in a hot suite, so the exposure—combined with the pain of the bite and the humiliation factor—was not a trivial event. I kept to my own procedures for rat handling and wonder if incidents like that didn't help me keep my edge. If you let your concentration lapse for even a second, any lab animal can turn a routine procedure into a dangerous accident.

Initially I did most of the work myself: injecting the animals, bleeding them, centrifuging and processing samples, personally assaying the virus during their infection, and conducting the neutralizing antibody responses. But by the time we were doing the last monkeys two or three years later, a lot of people were pitching in. John Morill and Jerry Jennings, both D.V.M./Ph.D. at USAMRIID, were leaders in many of our later experiments. It was very easy to inhibit the virus when we pretreated cell cultures or the monkeys with interferon, so interferon could be beneficial in preventing disease if someone was exposed to the virus. However, not unexpectedly, there was no effect if the interferon was given after the virus had already started to grow and spilled over into the monkey's bloodstream.

Sometimes you'd get useful data through highly irregular and unpredictable methodology. At one point I had a lab tech working for me named Mary, who happened to be a knockout. On a particular day, she was responsible for injecting some mice with the virus and she'd gotten Gil, one of the physicians who worked in the hot suites, to help her. Mary would reach over and get the mice out of the cage while Gil did the injections. Gil must have been watching Mary, because at some point, instead of injecting the mouse, he injected his thumb. Ironically, Gil's lapse in concentration moved the research along. He became our new laboratory test case.

He'd been vaccinated, with a neutralizing antibody response, so we didn't feel it necessary to put him in isolation in the slammer, but we watched him carefully, taking his temperature and analyzing his blood. He did have a spike in antibodies after the exposure but no disease and no viremia.

Actually, this is the way a lot of protective agents got human-tested at USAMRIID—through accidental lab exposure. It's not the same as doing a large-scale control test, but since you aren't purposely going to expose humans to untested stuff, this scenario, when it arises, can give you valuable data.

We tried to make it easy for other researchers to incorporate Rift Valley virus in their work. This led to a potentially important breakthrough in therapy. A guy named Ed Stephen came to work at USAMRIID and brought a then-new antiviral drug called ribavirin. He had been experimenting with it and it seemed to inhibit a lot of different viruses. We joined up with him to test whether it would work against Rift.

We started with mice and began with all conditions weighted to favor the drug. We gave the mice ribavirin first, then a small dose of virus. We gradually increased the dose of virus and began treating later and later, testing the limits. We found we could treat pretty far along in the course of the disease, say about three-fourths of the way through in the illness, and still show an improvement in mortality.

This kind of testing served several purposes. We wouldn't expect direct correlation between the results in mice and in humans, but success in mice gets you one step further. John Morill later did the same type of testing in monkeys and showed you could inhibit virus if you gave the drug within twenty-four hours of onset of viremia.

Time constraints for treatment were critical, we discovered, both from a medical and military standpoint. If someone zaps U.S. troops with Rift Valley fever, we need to know when to administer the drug for it to be effective.

By 1980, we knew ribavirin had potential for treating patients with Rift Valley hemorrhagic fever, but to date we've still not had a good chance to test it in humans. For one thing, it's hard to predict when the next epidemic of Rift will occur, and you'd need an epidemic to get enough patients for an effective study, since only 1 or 2 percent of the people who get Rift develop full-blown hemorrhagic fever. And as with most epidemics in faraway regions, if you wait for one to happen, it's usually peaked by the time you get over there.

The drug has tremendous potential, though, and we're still pursuing it today. CDC is putting together a protocol with the World Health Organization so that when a Rift epidemic next occurs, we'll be ready to go wherever it breaks out.

The final piece of the mystery was: where did it come from? Where was the virus hiding between epidemics? The answer came to us through a series of coincidences that brought a host (no pun intended) of talented specialists together.

Colonel Philip Russell at Walter Reed had an entomologist, Charlie Bailey, and a virologist, Joel Dalrymple—both protégés of his—whom he wanted to send to USAMRIID to work with us. He thought it would be good for them to learn about a different set of viruses and they could help us reach critical scientific mass. He was right on both counts.

We'd promised the USDA we would not let the virus loose and we planned to keep that promise. But containment worries increased a hundredfold when we started working with mosquitoes. The last thing we needed was one of these suckers infected with Rift flying away.

The blowers and negative airflow would help somewhat, since mosquitoes do only so well against air currents. We painted all the rooms inside the suite bright white so a mosquito on the wall would really stand out. Next, we hung nets throughout the suite, creating compartments you had to slip through to move around the suite.

One of the first things Charlie did was assess how much of a risk there was that local mosquitoes could be effective vectors if the disease were introduced in the United States. Sure enough, a selection of North American mosquitoes were quite efficient at transmitting the disease. While this proved again the importance of the work we were doing, it also made us feel even more paranoid about not letting this virus out of the lab.

We were also fortunate in knowing some really smart guys out in the field in Africa who coordinated and shared information with us. Glen Davies, a British veterinarian working in Kenya, and Bob Swanepoel, a British citizen born in Zimbabwe and working in South Africa, provided theories and data that helped the investigation tremendously.

Bob had studied virology at Edinburgh and got his Ph.D. and veterinary degree before he returned to Zimbabwe, to the veterinary lab there. As others had previously observed, Bob found that Rift was a big problem only in years when there was a lot of rain and mosquitoes were very dense. Where his thoughts differed, though, was in where the disease came from. A popular theory in Zimbabwe and in South Africa was that the reservoir was in forest such as the Natal Forest along the eastern coast of South Africa, from which it spread during rainy season throughout cattle-raising areas in the rest of the country.

This was not what Bob observed in Zimbabwe. He saw that in a rainy year he got samples of Rift Valley fever from cattle-growing areas all over the country, more or less at the same time. It seemed to erupt everywhere at once. He theorized that there was only one kind of reservoir, but there were multiple copies of it brought out by the heavy rains of epidemic years.

In trying to prove his theory, he plotted all his cases on an overlay that he placed on top of a map of the country. Bob noticed that the cases corresponded with a particular type of topography. The instances of disease were all located in or around vleis, an Afrikaans word referring to places where water pools near streams. The vleis, or the related structures called damboes in East Africa, have a specific soil type and geology such that they connect with the ground water. They don't always flood during rainy season because they will drain, but in years of heavy rain the ground water will rise and saturate them. In a dry season, these areas look like puddles of mud, with grass and sedges growing on the fringes.

Bob discussed his observations with Charlie, who immediately thought of research he'd done with another virus as part of the army's now defunct training program for field virologists. That project involved an arbovirus on the Eastern Shore of Maryland in the swamplands of the Delmarva Peninsula. A specific mosquito, *Aedes atlanticus,* harbored the virus and would lay its eggs in the salt marshes of the swamp. In heavy rains, the marshes would flood, the eggs would develop, and all the mosquitoes would come out at once. Many of them were already infected with the virus because the mother mosquitoes would transmit it to their ova.

The young mosquitoes would mature after hatching and then take a blood meal. While probing for a juicy capillary, they would deposit enough virus to infect the deer that provided dinner. The infected host would later pass on virus to other mosquitoes that fed on it but wouldn't be necessary for the survival of the virus. After a nice blood meal, the *Aedes atlanticus* seeks a quiet spot to deposit her eggs, already infected with the Keystone virus before they leave her body.

If this were so, it would represent a departure from the classic situation that was first elucidated by Walter Reed for yellow fever and which we thought took place when Rift Valley fever ravaged sheep and cattle herds. In the usual cycle, the virus is taken in by the mosquito with its blood meal and the mosquito undergoes a disseminated infection. After a few days, the virus has reached the salivary glands and when the mosquito bites another host she continues the chain of transmission.

Charlie reasoned that the vleis or damboes in Africa might serve the same purpose for some African mosquitoes as the salt marshes of the Eastern Shore. The pattern certainly fit the conditions Bob described. To test the theory, Charlie sent a young entomologist, Ken Linthicum, to work in Glen Davies' lab in Kenya and research the damboes there. After he arrived, they had a heavy, early rainy season in Kenya and he was able to gather mosquitoes from the water accumulating in the damboes and hatch the larvae in captivity, studying their development. Many had Rift Valley fever virus when they emerged, indicating they were infected at the time the eggs were laid. The eggs were very drought-resistant and could last at least several years—maybe more than a decade if they had to—waiting for a soaking rain.

With these studies, Charlie and his colleagues were able to prove that the maintenance of Rift Valley fever virus in Africa was related to a combination of weather conditions and the habits of several species of so-called floodwater *Aedes* mosquitoes. After the virus was rescued from the dried eggs, it could spread in typical arbovirus mosquito-host-mosquito-host fashion. Floodwater *Aedes* were the critical reservoirs, although a variety of mosquitoes could be vectors once they fed on a viremic animal.

This is a vivid example of how nature can replicate the same patterns using a different but equally specific set of reservoirs, viruses, and host organisms around the world. It also speaks to how much we can learn through surveillance of viruses and study of disease patterns.

This hypothesis explains why the virus died out in Egypt after the epidemic. After the Aswan Dam was constructed, there was no more alluvial flooding. There's also virtually no rain in Egypt, so there are very few flood-water mosquitoes. As sheep and cows died out or became immune in their recovery, it became impossible for the virus to survive simply by jumping from host to host. Without a flood-water mosquito, the virus can't maintain itself over the long haul.

There was a rather unpleasant epilogue to the story. By 1980, Rift Valley fever had essentially disappeared in Egypt and funding had dried up. One Egyptian lab announced it had uncovered the disease in two members of the Egyptian army, which it demonstrated through a hemagglutination inhibition (HI) test for Rift Valley fever virus antibodies. There was some concern at the Ministry of Health, which asked NAMRU-3 to confirm the finding. The virologist at NAMRU, Owen Wood, repeated the HI test and got the same results. But he decided to make sure by trying again with a fluorescent antibody test. He took infected cells and put them on a slide, fixed them with acetone, added the serum; if it had the right antibodies, it would adhere to the infected cells. Then he tried to detect the antibodies with fluorescent-tagged goat antihuman antibodies.

The test came out negative. Owen didn't understand it, so he ruminated on it for a while. He retested the sera from the soldiers, this time using fluorescent-tagged goat antirabbit detector antibodies. And lo and behold, he detected the presence of a control rabbit antiserum that we had distributed widely to Egyptian labs to help with their testing. Someone, in an attempt to make it look like Rift was still circulating so they could keep their funding and look important, had spiked the convalescent samples with the rabbit serum. So we must always remember that science is subject not only to the vagaries of nature but also to the frailties of human nature as well.

The Rift Valley fever story is a saga of scientific suspense and great virology lessons leading to a positive conclusion. We're not always that fortunate. This time, we found out why the outbreaks occur where they do and we can now even predict them to some degree through satellite remote sensing. We have a vaccine proven to protect against the two versions of the virus we know about, and we've begun setting things up so that when it hits again, we have an experimental drug protocol ready to go to the most severely ill patients. The answers don't always come quick and they almost never come cheap, but if they translate into lives saved and countless people spared suffering, then I'd say it's well worth the effort.

7

The Education
of a Military
Physician

THERE ARE MANY true stories of great leaps forward—what we now call paradigm shifts—and heroic events in the annals of medicine and public health. But two are particularly important in understanding the many changes taking place in our approach to infectious diseases today: those of Drs. John Snow and Walter Reed.

In 1854, London was in the grips of a cholera epidemic that had killed more than 500 persons. Cholera is an incredible scourge even today, a pestilence that causes violent and uncontrolled diarrhea, which basically drains its victims of bodily fluids, ultimately leading to shock and often death. Snow, a physician of many accomplishments, used sound epidemiology to identify a particular pump on Broad Street in the Soho district of London as the source of the disease. He traced the feed of the pump and found that it was supplied by water from the lower Thames, which had been contaminated by sewage from farther up the river. The water systems drawing from upriver were not associated with disease. As the now legendary story goes, Snow convinced authorities to remove the handle from the Broad Street pump and, in so doing, ended the epidemic.

This pure epidemiological approach did not presuppose the existence of any infectious organism. Indeed, the bacterium that causes cholera had not even been isolated at the time. Pure deduction provided the source of the problem and also a solution without intervening details on the causation of the disease.

This is in many ways similar to the cigarette story today: the epidemiological links to disease are clear although the mechanisms are not well understood. Changes in the demographics of cigarette smoking have reliably led to changes in the incidence of lung cancer in the United States. Men, as a group, are smoking less than before and their incidence of lung cancer is down. Women, as a group, are smoking more and their incidence is up.

Walter Reed's story is a very different kettle of fish. Yellow fever had been a savage disease in Africa for centuries before it was brought to the New World with its vector mosquito, *Aedes aegypti*. In 1648, *xekik*, or "bloody vomit" in Maya, struck in the Yucatán, representing the first major yellow fever epidemic recorded in the Americas. The disease established itself so strongly in our hemisphere that it continued to devastate settlements and change the course of history. In 1702, the fever claimed one in ten New Yorkers. In 1793, 15 percent of Philadelphians were stricken. In 1802, the French lost their general and so many of their 25,000 troops in Haiti, that the slave revolt led to the island's independence two years later. The disease spread to all the great cities of the Americas and even temporarily invaded England, France, and Spain. One of the centers of dissemination was Cuba. Continuous transmission of the disease often resulted when ships stopping in Havana departed with a lethal gift for their next port of call.

Arguments raged over the cause of the disease, and it was attributed to contagion, miasmas, and a variety of other mechanisms. Viruses were unknown and arthropod transmission of diseases was not suspected for any illness.

After the United States occupied Cuba in 1899 following the Spanish-American War, yellow fever casualties among the garrison threatened to lead to the loss of the island. In addition, control of the disease could safeguard against its importation to New Orleans and other coastal and riverine ports of the southern United States. Walter Reed, an army surgeon in his forties, was sent to organize a commission and to save the day for the home team.

Reed, like Snow, had what they would have called a "disease" to study, but really it was nothing more than what today we would call a "case definition," since neither had any way to prove that their cases truly had the disease and not another condition resembling it. Unlike Snow, Reed was a microbiologist as well as a physician and epidemiologist. When he arrived in Cuba he was confronted by a dismaying array of confusing epidemiological information and unproven theories. Among those theories were those of Carlos Finlay, a brilliant sixty-seven-year-old Cuban physician who had been interested in yellow fever for most of his professional life and had written a paper almost twenty years before suggesting that the disease was transmitted by mosquitoes.

Reed was intrigued by Finlay's reasoning, which had even reached the point of suggesting *Aedes aegypti* as the responsible mosquito. But the theory was incomplete because Finlay seemed to regard the mosquito as a miniature syringe that moved the infective agent from one person to another. Furthermore, Finlay had been unable to demonstrate mosquito transmission experimentally.

However, Reed also had the acquaintance of Henry Carter, an officer of the Marine Hospital Service (later to become the Public Health Service). Carter had studied a series of small epidemics on farms along the Mississippi. After the first case or index case returned from an epidemic area to his home, two weeks elapsed before new cases appeared. Carter was struck by this because he knew that people coming into a yellow fever epidemic needed less than a week to get sick. He called this extra time the "extrinsic incubation period," to contrast it with the "intrinsic incubation period" in the patient himself. The extrinsic incubation period was profoundly influential in Reed's decision to pursue the mosquito idea and also provided the clue for the success of the experiments—the mosquitoes had to be held for several days to allow the virus to develop in them before they could transmit. Also at that time, a British army physician by the name of Ross had described the mosquito as a vector of malaria—but the malarial parasite was large and visible under the microscope and not an invisible killer like the yellow fever virus.

Reed's experiments were brilliantly done and showed that there was no contagion or miasma, that the mosquito could transmit yellow fever after an extrinsic incubation period of five to twelve days, and that the agent of yellow fever could pass very tiny pores in filters and so was not a parasite or bacterium but what would later come to be known as a virus. Thus, Reed and his group established the first filterable human virus, the first arthropod-transmitted virus, and the first existence of free virus circulating in human blood.

It remained to Major William C. Gorgas to launch a massive eradication campaign to free Cuba of *Aedes aegypti* and prove beyond any
shadow of doubt that the mosquito transmitted yellow fever and that
yellow fever could be controlled through striking at the mosquito. He
translated science into public health so effectively because the disciplined teams of a military occupation could attack the well-defined
breeding sites of *Aedes aegypti* and eliminate them. Gorgas later
repeated his feat in Panama and made it possible for the United States
to build the canal where the French had failed, both from a poor
design and from the ravages of yellow fever. The hospital in Panama
where I both worked and was treated was named after him. Gorgas
was so successful that by 1902 there were no new cases of yellow
fever in Havana for the first time in fifty years! New Orleans, however, got careless, and in 1905 they had another epidemic of 3,400
cases, with 452 people dying, to add to the yellow fever cemetery on
Rampart Street.

Notice the difference between Snow's and Reed's approaches. This
contrast between the classical epidemiological study that yields a
practical solution and research that leverages the microbiological
findings to provide an equally compelling resolution to a problem
that had stumped the best minds of the day in infectious disease for
centuries is an exact parallel to the two basic points of view in infectious disease epidemiology. The type of epidemiological study Snow
used is still uniquely useful in some settings, but Reed's approach is
ultimately so much more powerful; although based on epidemiology,
it can provide the causative organism, which then becomes the focus
of efforts to control the virus, allowing us to use much more subtle,
specific, and sensitive tools to defend the public health.

Reed, of course, is himself known today for the major army hospital named for him in Washington, D.C. But Reed was more than that:
an exemplary guy, a brilliant organizer and medical analyst who led
a tremendously talented and dedicated team and then gave credit
freely to each member of that team, which accomplished such a great
deal in so short a time period.

I used to wonder why we never heard any more from Reed after
this tremendous feat until I looked it up and found he died suddenly
of appendicitis in Washington, D.C., on November 23, 1902. Like
John Snow, who died only a few years after his own historic contribution, Reed helped untold millions of people, only to succumb
shortly thereafter to an unrelated disease.

But the ideas themselves live on. The concept of extrinsic and intrinsic incubation weighed in our studies of the role of floodwater mosquitoes in the transmission of Rift Valley fever, for example, and has tremendous applications to many virus diseases today.

At USAMRIID we were interested in evaluating and finding defenses against anything that foreign countries might weaponize or apply to biological warfare. This was a bit like looking in the hall of mirrors. If "they" were working on it, did that mean they were weaponizing it or just publishing a paper or two to keep us on our toes? Should we work on the agents they published on, or the ones they *didn't* publish on? With the Sverdlovsk incident in 1979, we found out that the Soviets had a major program weaponizing anthrax, but it wasn't until the fences came down that they published their filovirus work in the open literature and we knew they were toiling in that particular cesspool of horror.

One of our key missions became research into antiviral drugs. This would have applications to both military and civilian public health, but Colonel Dick Barquist was especially determined that we find some way to treat each of the viruses we had people working on in the lab.

Barquist was truly an extraordinary CO. He cared about all the programs and always kept abreast of what was going on under his command. He made it his business to know everybody in every department, their strengths and weaknesses. The only complaint I ever had was that he could be slow to make decisions. But he had excellent judgment and could pull off a fast, wise one when the situation called for it.

Pleased with the potential ribavirin had shown against Rift Valley fever, we wanted to see if it would be similarly effective against some of the other hot viruses. Machupo, with its high mortality rate, was high on our list, especially since we knew the Russians had been working with it.

Gerry Eddy had been talking with Karl Johnson at CDC about his successes using ribavirin against Machupo, and Karl wondered if it would also be useful against Lassa fever, an often fatal disease of annual epidemic proportions in West Africa. Lassa was caused by an arenavirus that chronically infected rodents, and was believed to be spread by the rodent reservoir as well as through person-to-person transmission. Finding ways to prevent and/or treat Lassa was a high priority for CDC, and Karl had a project team out in Sierra Leone to study it.

Ribavirin was not without its problems, however. Few drugs are. The company hired by the manufacturers of ribavirin to do all the long-term testing (including toxicology) was the same company responsible for testing red dye no. 5. As anyone who remembers the original red M&M's will tell you, the dye was eventually pulled as being potentially unsafe. When the antiviral crew looked at the ribavirin data from this company, just as with red dye no. 5, it didn't hold up: audits were not done right, animals were missing. In initial safety tests, any irregularities at all make everything suspect.

My section, especially Peter Jahrling, who did most of the early ribavirin work with Lassa infected monkeys, delved into treatment of monkeys with convalescent plasma and ribavirin, and eventually I had my next six-month review with Barquist. I presented the data we had on Lassa so far, showing that antibodies from convalescent plasma didn't work, but we were beginning to test with ribavirin. Barquist considered everything carefully before he reacted to my summary. I can still see him putting his pen down to look up and ask, "Are you telling me that I have people here working with a potentially deadly virus and I don't have anything to offer them if they get infected in the laboratory?"

"Yes, sir, that's right," I gulped in response.

"Well, then," he said, "you stop any other experiments you're doing and find out what we can do. I want that worked out first thing." Barquist was worried about the big picture, but he was first and foremost going to take care of his people.

CDC was doing a lot of work on Lassa and needed help with research and funding. It served both our interests (and those of the taxpayers who supported us) to collaborate, especially if it led to proof that a new antiviral drug was safe and effective against a variety of diseases. So after we did enough toxicology testing to have confidence in ribavirin's safety and probable efficacy, Karl Johnson picked it up to start human trials for efficacy in Sierra Leone. In the meantime, Peter continued to work on antibody testing in monkeys.

We discovered that Lassa fever was unlike most of the other viruses we'd worked with. With Rift Valley or Machupo, for example, you develop neutralizing antibodies about the time you recover from infection. These neutralizing antibodies will protect if you put them into an animal with no antibodies and challenge (inject) them with virus. The differences with Lassa began with the fact that you don't develop neutralizing antibodies until three to six months after recovery.

Another difference lay in the dose for therapy. To be effective, antibodies had to be given in much larger quantities than for other viruses. And not all convalescent sera with FAs (fluorescent antibodies) would work. We had to develop an in vitro test to determine which antibodies would protect.

Back at CDC, Karl had set up a prominent and accomplished young epidemiologist named Joe McCormick in Sierra Leone to develop the field station and oversee oral ribavirin trials. I remember thinking: *You get to work with really sharp people. This guy is really interesting, very smart, very charming.*

Anyone who knows or has worked with both of us knows our relationship did not stay so warm and friendly. In terms of personality, I think we've always taken different approaches to things, as would become clear when we squared off over the Ebola outbreak in Reston, Virginia. But we both have had the same job, the one Karl Johnson originated at CDC, and I think we are both deeply dedicated to public health.

The oral delivery of ribavirin that Joe was going to test might not have been anyone's first choice, but an intravenous trial was prohibitive logistically and for reasons of cost. Even if needles hadn't represented such a hazard, giving IV infusions is not a routine procedure in Africa. You need people experienced in setting them up and keeping the line open. In addition, the price of a bottle of IV fluids, an IV administration set, and a disposable needle is a huge factor in the cost of hospitalization of a patient in the Third World.

The government of Sierra Leone imposed additional strictures on the study by not allowing a randomized double-blind control trial—a study in which patients who would receive the drug are selected randomly and neither the patients nor the doctors know who's getting the drug and who's getting a placebo. If a drug works, obviously you'd want everyone to receive it, but the only way to confirm that it works (and is safe) is to use a double-blind control. Historically, the shortsightedness of trying to rush a drug to market before good data are in has been proven time and time again, but there is always pressure from those who say, "If you think it works, give it to everybody."

We have seen this happen with various AIDS trials. There was such a clamor over AZT that the trials were stopped after what we call surrogate markers were reached. That is, there was evidence of clinical improvement and improvement in neurological function, so people wanted the drug out there. The Europeans later ran a trial through to the end point which showed AZT does not prolong life. Not only did the aborted trials provide false hope and unreasonable expectations; they also hindered physicians treating patients with AIDS. If they knew from a controlled, completed trial how and when resistance to AZT appeared, they could have adopted a completely different strategy using the drug. Instead of prescribing it early on in the hopes of prolonging life, which enables resistance to set in before the drug has a chance to treat specific symptoms, they'd wait until it would be able to relieve suffering.

In Sierra Leone, Karl and Joe had to judge ribavirin efficacy using what we call retrospective controls. They compared the people receiving ribavirin in the study with people who had had Lassa fever in the past, by using detailed questionnaires and reviewing blood samples and charts. In short, they took the most scientific approach they could, given the constraints placed on the study.

Retrospective studies are difficult and the data they yield has to be viewed with a dose of skepticism because there may be conditions that make present cases incomparable in some way to the previous cases. There's always the possibility that when you give the drug you unconsciously give better supportive care overall, for example, or that people in local villages will hear about the new treatment and so everyone—including mild cases—will come to the hospital and the drug will appear to work. Even less subtle changes can skew the data: maybe a main road becomes impassable during rainy season, so you don't see any patients from a particular area of the country.

Karl and Joe ran the study for two months. When they compared the course of disease and mortality in untreated individuals with those of patients who received two grams a day of ribavirin orally, they found no significant difference. This apparent ineffectiveness was partly explained by our research at USAMRIID. Our studies and some of the data on pharmacokinetics indicated the oral drug was absorbed relatively slowly and incompletely. We thought they could get the blood levels up more quickly and give the drug a better chance if they used an intravenous drug. We modeled an IV regimen in monkeys.

Although encouraged by our work, Joe wasn't interested in running an all-out IV trial on humans in Sierra Leone. There were arguments for and against a trial using IV ribavirin and both made sense, depending on your outlook. In Sierra Leone, Joe's argument was that he couldn't justify using valuable funding on a therapy that—even if proved useful—wouldn't be usable there because African hospitals couldn't use or afford IV sets. Fair enough.

The attitude the army took is that they had IV sets and they didn't care. My feeling was that first you give a drug every chance to show it's effective and then work toward practicality in the future. It is always easier to bet against a drug, because in the majority of cases a new drug will not live up to its promise. Barquist didn't mind paying for it because it was data worth having from our standpoint.

Karl, analyzing the data from the first set of controls, made an interesting discovery. He found that you could predict the mortality of Lassa fever from measurements you made when the patient walked in the door. There were two important predictors. One was the level of a certain enzyme, known as AST, which is often elevated in cases with liver and muscle damage. If a Lassa fever patient came in with elevated AST levels, there was an increased probability he or she would die. Similarly, the level of virus in the blood gave an indication of whether that patient would make it.

Karl got this new data out to Sierra Leone so they could target the studies for patients who were most likely to die. This represented a tremendous step forward in screening. When they started eliminating low-risk cases from the trial, it turned out that both oral and IV ribavirin were useful, although the IV preparation was more effective.

Shortly after this discovery, Karl left CDC. But his efforts laid the groundwork for much of what we understand about Lassa fever today. Joe took over as head of Special Pathogens and began dose-seeking, comparing the efficacy of the regular dose of IV ribavirin with half that amount to see if you could get the same benefits while using less, which would save money and minimize toxicity. With Joe's departure from Sierra Leone, those trials were not closely monitored, however, and the study was not randomized. If someone went to the hospital and was judged to be really sick, the locals might give him a larger dose, for example, not realizing they were skewing results and could harm the patient. If the drug was in short supply, all patients might get only half a dose. Efforts trailed off, and when I got to CDC a few years later and tried to reestablish the study, it ran successfully for six months, until Liberian rebels looted and burned the hospital down, killed a Dutch doctor, his wife, his daughter, and an elderly priest, and ran everybody off. Increasingly, efforts to combat disease in the Third World, particularly in Africa, are thwarted by internal political violence.

Although we had been talking about studying Lassa for a long time at USAMRIID, we really didn't get a program going until we started coordinating with CDC. Peter Jahrling, for one, was hot to get into it after his work on Venezuelan equine encephalitis. But as so often happens in a bureaucracy, even if everybody agrees something is useful and important, there are a number of logistical elements that have to be wrestled into submission, keeping progress at bay. Space had been assigned for the Lassa project, but there was someone already working in it and he had to finish one more experiment and didn't have a place to move his incubator. Then we had to wait for the next meeting of the capital equipment committee to get funds allocated for additional equipment. Based on the normal time frame for establishing this type of project, we were probably still about a year away from getting started when CDC forced our hand through an act of negative providence, one of those accidents, if you will, so often responsible for moving science forward. The folks at CDC contacted Gerry for help when two of their guys had a lab exposure to Lassa. British epidemiologist Dick Keenleyside and American lab technician Mike Dudley were exposed in the CDC lab in what we in the scientific community refer to as a TFU: a total fuck-up. The CDC wanted to book a couple of rooms in the slammer.

Keenleyside had been in Sierra Leone leading a study on Lassa transmission in households. It involved visiting the home of a Lassa patient (the index household) and trapping rodents there and in two other houses, a near control and a far control. He bled people in those households to see if they had antibodies for Lassa and tested the rodents for virus. Everyone was then retested at intervals to check for person-to-person transmission in the houses and to see if there were any critical differences in the number of infected rodents in the homes.

Keenleyside had returned to CDC with a bunch of sera from Sierra Leone (from people and rats) that all needed to be tested. Dudley was to do the actual virus isolation and antibody testing, but Keenleyside was going to help Dudley get set up.

Whenever you do an analysis on a laboratory accident, you often find there were several factors that contributed to its occurrence. What happened to Keenleyside and Dudley, though, is a perfect illustration of why you need both proper facilities and appropriately trained personnel to do lab work with biohazardous agents.

The main problem was that Keenleyside had never worked in a

lab. Given the dangerous agents we work with, the people in the labs must be experienced.

The guys trapping the rodents in Africa didn't take any precautions; they had all had Lassa fever before and were naturally immune. Similarly, at the local lab most of the workers processing samples were Lassa-immune. Keenleyside simply was not used to working with a hot virus in containment.

The samples were being stored in a freezer in the hall, just outside the new Level 4 lab, at Level 0. At that time, CDC had a small BSL 4 lab up and running. There still wasn't a lot of room to work with, but protocol would have been to sort those samples in the lab.

At both CDC and USAMRIID, hazardous agents are stored in ultra-low-temperature freezers called Revcos, which look more like coolers than the refrigerators we use at home. They open from the top, not from the front, which guarantees you never have the unpleasant surprise of having a frozen biohazard fall out on your foot. You do, however, trade off the convenience of frost-free storage for safety.

The samples in the Revco were in plastic test tubes held upright in a little test-tube rack. They removed the frigid test-tube rack and placed it in an ice tray atop the Revco, which, like many appliances, was warm on the outside and began to defrost the samples.

What they planned to do was have one read off the numbers on the side of the tubes for the other to jot down, so they'd have a record of which samples were being worked up. Dudley was pulling the tubes out of the rack one at a time and got to one that was stuck, frozen to the rack. He grasped it on top and pulled harder until they heard a little pop as the cap came off in his hand. Keenleyside felt something moist on his face, noticed several tiny spots of blood on Dudley's shirt, and realized what had happened. They talked to Patricia and Joe, and at first Joe decided just to watch them for three or four days, taking their temperature, but then decided he'd like to have them in isolation—just in case.

We had previously made a formal agreement with CDC promising to support them if anything like this happened. They didn't have a slammer and certainly neither did any public hospitals.

We turned down the sheets at our facilities; it was Joe's responsibility to get them to us. The guys were not sick; the disease was still in the incubation period. We arranged for an army plane to fly them to Hagertown and then took them to USAMRIID wearing respirators rigged to filter the exhaled air rather than the inspired air. This may have seemed like overkill, but Col. Barquist was a belt and suspenders man. Today, increasing evidence has shown that hemorrhagic fever patients are not infectious until they become clinically ill, if then. In any event, they made it to Frederick safely and we locked them in the slammer.

It was no small disruption to life at USAMRIID to care for them. For one thing, it takes serious commitment of personnel to run the slammer. Whoever had expertise with the virus in question had to put all other work on hold to test patient samples. Then we had to keep all other viruses out of the lab to avoid possible contamination, so nothing else got done. The saving grace was that at least these guys came from outside our lab. If they were our own employees it would have meant two workers out of commission from the select group who would have been capable of testing samples and doing the rest of the work with the virus.

We also had a team of nurses and corpsmen specifically to operate the slammer. When someone went in, they finally got to work with live ammunition, so to speak.

One of the biggest problems you face managing hemorrhagic fever patients in a modern hospital is getting their samples worked up, but USAMRIID was perfectly set up to handle this. The building was designed with a dedicated clinical laboratory, composed of two virtually identical modules, identically equipped. One was a Biosafety Level 4 laboratory; the other was a regular clinical laboratory. The regular lab kept busy most of the time doing tests on dependents who came into a clinic staffed from nearby Walter Reed or Fort Meade. Most workers were regular clinical lab techs, but some were trained in BSL 4 work and assigned to the containment lab.

With just one day's notice, we were able to get all the space, personnel, and equipment we needed ready ahead of time. For example, we didn't have a fluorescent microscope (needed for fluorescent antibody testing) in the hot suite that was to be dedicated to the virus. You can't just take equipment out of the Machupo lab, for example, and use it in a Lassa lab. Part of the containment measures was that each lab was 100 percent independently functional. Barquist got the army to buy a piece of capital equipment overnight—about $20,000 worth in 1980—no small feat. We were up and running before our patients even arrived.

As we worked to get the facilities ready, we also had to figure out some strategy for possible treatment. The accident happened at the front end of studies in Sierra Leone, before ribavirin was ready for

use in humans. At the time, we didn't know much about treatment of Lassa fever outside preliminary experiments in guinea pigs and monkeys with convalescent plasma. The theory was that if you took plasma from people who had recovered from the virus and gave it to people who had never had the disease, the antibodies in the plasma would protect them. This treatment had worked well with other viruses, such as Argentine hemorrhagic fever virus. We hoped to provide a ready-made immune response.

We happened to have about forty units of convalescent plasma on hand for just such an occasion. A Peace Corps volunteer who had Lassa fever and survived allowed us to repeatedly draw plasma from her. She'd returned to the United States and was donating plasma in a Red Cross and American Association of Blood Banks-certified facility at the rate of two units twice a week. It had been tested for everything known at the time and found to be safe, and it contained the all-important Lassa antibodies.

We explained to Keenleyside and Dudley that we didn't know if it would be helpful or not. Both agreed that they wanted some. We gave them the plasma and watched them closely for several weeks. As it turned out, while both had passive antibodies from the plasma, they never developed an active antibody response, never got sick, and we never cultured any virus out of them. We kept them in the slammer for over three weeks (the outside edge of the incubation period for Lassa plus six days—the time it took to run cultures for Lassa).

The experience was important on several levels. First, we were surprised they didn't get infected. Back in Atlanta, Patricia Webb tested the blood in the tube they were exposed to and it was positive. There were thousands of virus particles per milliliter of serum. We didn't know if Lassa was less infectious than we thought, or whether this was a less infectious strain, since we didn't know how many strains of Lassa are out there. It could also have been that they were more resistant as hosts than the people who get Lassa fever; or maybe the convalescent plasma we gave them helped.

It was interesting to see how the two guys responded to being in the slammer. Keenleyside was nervous and concerned the whole time. In Sierra Leone, he'd seen people die from Lassa and was extremely upset to think of that happening to him. And while the slammer offered state-of-the-art facilities and specially trained personnel, these guys didn't know any of us. Except for phone calls, they were completely cut off from their loved ones and the outside world.

Keenleyside was also beating himself up for haphazardly handling biohazardous materials in a hallway: no one wants to get critically ill or die because they made a stupid mistake.

Dudley, in contrast, was very low-maintenance and completely matter-of-fact. Perhaps it was easier for him to deal with because the virus was more abstract to him—he'd never seen the ravages of Lassa fever up close. As long as the results from the laboratory tests came back right for him, he understood he was okay. And each day that passed made it only more likely the numbers would be in his favor again the next day.

The experience crystallized our plans on how to handle exposures. We were in a sensitive situation at USAMRIID. The people of Frederick loved Fort Detrick on payday. But whenever something happened, it became "Fort Doom" all over again and questions arose about how well they could trust the army.

It wasn't like we were working in Sierra Leone, where Lassa was everywhere. We brought this and the other diseases to Frederick. Nor was it like CDC, where the viruses were being worked on in the general national interest. We looked out for the United States too, but our job was foremost a military operation: "Research for the Soldier." And in 1979 and 1980, post-Vietnam trust in the military was pretty low.

The first guy to get slammed at USAMRIID was my good friend Ralph Kuehne. Gerry Eddy, recently appointed head of the virology division, had put Ralph in charge of some monkey experiments with Machupo virus. Ralph wanted to determine how many red blood cells were in Machupo-infected monkey blood. If you centrifuge blood in a small tube, the components separate out and it's easy to measure what proportion is red blood cells. Ralph was doing this procedure (called a microhematocrit, and no longer done this way because of his exposure) using tiny glass tubes. Typically, you seal a tube by plugging up one end with your finger and sticking the other end in some clay. Ralph was working in a hood line and got up to the point where he stuck the tube in the clay. Sense of touch is greatly diminished in the hood and it's difficult to tell how hard you're pressing on things. Ralph pushed a little too hard and the tube broke and cut right through his glove into his index finger.

Ralph immediately recognized this as an "Oh shit!" moment. He yanked his hand out of the glove, squeezed out some blood and quickly disinfected the finger. But the presumption had to be that since he was doing this at a time when the monkeys were viremic, this was a very serious exposure. As we saw with poor Donato Aguilar, direct blood-to-blood exposure is about as serious as it gets.

In field work I've heard people speculate that after being exposed you could avoid infection by amputating the offending digit or limb. Would it have saved Donato, for example, if instead of merely squeezing out blood and disinfecting the wound he'd actually lopped off the finger? During the Rift research I saw an experiment that dramatically illustrated why amputation wasn't necessarily helpful.

I'd asked entomologist Mike Turell whether mosquitoes actually inoculated Rift Valley fever virus directly into the tiny capillary they drank from and therefore right into the bloodstream, or whether they deposited the virus in the surrounding tissues. I knew if it went directly into the blood it would be like receiving a drug (or toxin) intravenously, which would be far faster and more efficient than hitting tissue. Mike devised an experiment in which he used newborn mice (highly susceptible to Rift) and mosquitoes infected with the virus and created a shield so a mosquito could bite a mouse only on its tail. At determined time intervals, Mike then cut the tail off.

Results were mixed as far as answering my question went. After some mosquitoes took a blood meal, even if the tail was removed immediately, the mouse got sick and died. In other cases, Mike left the tail intact up to five minutes and the mouse survived. It seemed some mosquitoes hit the bloodstream but others left the virus outside the vascular system. If they got to blood directly, it moved much too fast to be stopped by amputation. Only with a subcutaneous inoculation did they have a shot.

In a human being it takes only a few seconds for blood from a capillary to reach a vein; so in Ralph's case, Machupo could have been well downstream before Ralph even pulled his hand out of the glove. Ironically, Ralph's exposure happened as he was experimenting with the usefulness of immunoglobulin as treatment for Bolivian hemorrhagic fever. Karl had sent human immunoglobulin prepared from convalescent patients to try out on monkeys. If it seemed effective they'd move to human-use trials in the field.

Ralph obviously sped up the plans quite a bit, as they immediately jumped to testing therapeutic measures in humans—on him. Before he had a chance to get comfortable in the slammer, Ralph got a shot of Machupo antibodies. Without tests, they didn't know the correct dosage, so they probably started out with the scientific standard in a case like this: a swag, or silly, wild-ass guess, delivered with crossed fingers.

They also decided to give him a shot at the site of inoculation, since there was bound to be a store of virus locally. This sounds like a great idea until you think of how sensitive the end of your index finger is. This is where you have the most nerve endings per square millimeter: more than the tip of your nose, the bottom of your feet, and—Ralph swears this is true—more than the end of a certain male organ considerably less likely to suffer accidental exposure in a laboratory.

Charlie White went in to see Ralph and warned, "Ralph, this is gonna hurt, but I have to put some of this into your finger." They could have anesthetized him with a finger block, but in an emergency situation you don't have time to get an experienced person to do it; you're trying to act as quickly as possible. Charlie took a syringe full of immunoglobulin with a fine-gauge needle, stuck it in Ralph's already punctured finger, and squeezed in as much as he could. Ralph is a tough guy, but he told me he couldn't stop the tears in his eyes. The pressure from the needle and all that fluid going in his finger was the most excruciating pain he'd ever felt in his life.

They kept Ralph in the slammer for twenty-one days. He was never symptomatic, he never bumped his antibodies, and they never isolated virus in his blood. But getting direct blood inoculation of Machupo virus (presumably at high titer) and not getting sick is highly unlikely. I believe Ralph was infected and the immunoglobulin he received neutralized the virus. Bolivian hemorrhagic fever causes death in one-quarter of the people who develop infection via any route, and Ralph was exposed the same way as Donato Aguilar. Ralph definitely dodged a bullet.

After he got out of the slammer, Ralph received something of a longer-term sentence. I'm not sure if it was meant as a warning or a joke or what, but they made Ralph safety officer. I guess in addition to being a smart guy and talented microbiologist, Ralph possessed the most relevant experience for the position.

Gerry Eddy had had his own lab accident, which happened shortly after I got to USAMRIID. George Frye, whom we had hired during the shakedown cruise of this hot suite in 1971, was still on board. He was taking Machupo samples out of the hood line. The procedure (the SOP I set up when I was at USAMRIID) was that virus samples went into vials, which were brought into an air lock in the hood line and sprayed down with Lysol. The vials were screw-capped. After the spray, they sat in the UV air lock for fifteen minutes for further disinfecting. Then you'd take them out of the air lock and put them in a plastic bag for placement in the freezer.

George was taking some of these vials to the freezer and one tilted over. In spite of the fact that the top had been screwed shut, it leaked. (Following this episode, the procedures were rewritten so that the vials would be sealed in something else before passing through the air lock: either another screw-top vial or a sealed plastic bag.)

A few drops leaked out of the vial and fell on the hard surface of the Revco. We had to assume the possibility of aerosol generation. Since the Revco is about chest high, any aerosols generated would have been pretty close to George's nose. George, who was the only one in the suite at the time, picked up the phone and called Gerry, who ran down, changed into a scrub suit, and rushed in there without putting on a respirator. Again, in the heat of the moment, you can be so focused on the human details you forget the prmary concern. He had a much lower risk of aerosal exposure than George (especially with the blowers constantly sucking out air), but he was still at risk.

Gerry and George showered and we had a conference in Barquist's office. Gerry was at much smaller risk than George, but it would have looked funny if we put a technician in the slammer and let his supervisor off the hook. Gerry was in an uncomfortable position at the meeting, too, because he attended wearing three hats: he was exposed; he was supervisor to one of the exposed; and he was the most knowledgeable person on the virus.

Barquist concluded, "I think it's obvious you should both be in the slammer."

We discussed therapy options and decided not to treat either of them. We thought that the aerosols generated represented a significant risk but that they probably weren't enough to infect them and we planned to monitor their blood and temperature closely. If we got even the slightest warning sign we could always intervene later.

At that point, Gerry tried to bargain for his freedom: "George didn't have enough exposure to justify treatment, and I've had even less exposure. Do you think I really need to be in the slammer for three weeks? I've got a lot of stuff to do."

Barquist stared him down and said in an even but definitive tone, "I think you should both be in the slammer." We called in George, explained the decision to him, and put them in the slammer.

They each got their own room, although they were adjacent, so the heads of their beds were up against the same solid, blank wall. The ventilation from the two rooms met on the way to the HEPA filters that led outside, so if the place ever went ambient there might have been some risk of cross-contamination, but that risk was very slight.

Right from the beginning it was clear we were dealing with opposite patient types. Gerry is a large, imposing guy who's wound like a spring. He's very intense and active and always has a million irons in the fire. At this point, he was heading the virology division, which was a very visible, fast-growing group that he couldn't afford to leave for three weeks. Outside the lab, he also had a wife and two kids.

George is a lab tech who's reliable and hardworking but certainly less intense over missing time at work than Gerry. While Gerry bounced off the walls in his side of the slammer, George watched TV, smoked his pipe, and enjoyed the quiet. We sent in a paint-by-numbers kit and George took up painting. For him, this was leave with pay that didn't use up any vacation days!

The paint set didn't interest Gerry, so the head of bacteriology, Ken Hedlund, sent in some rum he picked up on vacation in St. Croix, but it became apparent that nothing short of tranquilizer darts was really going to help. Then Gerry started doing chin-ups on the shower bar in his suite and suffered chest pains.

His blood samples, like George's, had given us no cause to worry that he was infected with BHF, but chest pains are nothing to ignore. We had a dedicated portable X-ray machine in the suite and, wearing our space suits, we took a full set of X-rays. The film cassettes stayed in a plastic bag, which didn't interfere with the image but could be dipped in disinfectant and passed out of the suite to be developed like any other X-ray.

Gerry had a pleural effusion and a slight fever and we didn't know why. We were well into the third week in containment, the absolute outside of incubation for Machupo. It seemed especially unlikely that Gerry was infected, since George was the one with the greater exposure and he was fine.

We decided to let Gerry out a few days early to see his regular physician. He was worked up from head to toe and was never officially diagnosed, although there was speculation that he had an atypical reaction to a vaccine he received before he went in.

Gerry's case was the most frightening experience I had while at USAMRIID, which was ironic in view of his low risk of exposure. In all the time I was there, we never had a truly symptomatic lab exposure. In most cases I'm sure it was because they weren't infected, but in others it is likely they were infected but the treatment they received thwarted the virus before they developed a significant immune response.

As Gerry's virology empire grew, things were going well for me professionally as well. By the end of 1979 I'd been promoted to Lieutenant Colonel and was made chief of the Department of Viral Pathogenesis and Immunology the next year. As Gerry had predicted, I was rapidly garnering more and more costume jewelry for the uniform I never wore with much comfort or style. Some people are made for military uniforms; I was made for sport shirts, blue jeans, and sandals.

In 1981 I won the Surgeon General's Award for my work on Rift Valley fever. I was also happy in my personal life. With Susan I had the best relationship of my life. Mayra usually spent the summers with us, but one year her mother went to Panama and she stayed a whole school semester and summer with me. By then she'd already spent enough time in Frederick to have friends of her own as well.

Everything was moving along well except the actual divorce. It took a long time to finalize, which was frustrating because I'd emotionally felt divorced for a long time. But our unofficial arrangements regarding child support and shared custody held. As long as I sent money, I could see Mayra on a regular basis in D.C., and Antígona and CheChe on rare visits to California. In December 1981 the divorce was finally made official, and on May 14, 1985, Susan and I were married. On my third try, I finally got it right.

It's always been interesting to me to encounter the extremes in reactions to the concept of BSL 4 work. There are a lot of people who think we're crazy for wanting to work with this stuff; it can get pretty hairy, particularly when we go out in the field in Third World countries to study diseases for which we have no known cure or treatment. I know in my heart that while I've always been aware of the dangers, my overwhelming interest in the science far outweighed concerns for my safety—often to the chagrin of my loved ones. I've tried my best not to take unnecessary chances and have applied my fatalism to my work. The best I can do is try to be prepared and do my job diligently; if something is going to happen, it's going to happen.

As committed as I am to science, I've never been one to invite trouble. In the 1980s, though, I had the privilege of working with an incredibly gifted researcher and virologist who did just that. Argentinian scientist Julio Barrera Oro had challenged all bounds of scientific devotion (and perhaps good common sense) in the name of scientific truth, injecting himself with a virus with a high mortality rate, essentially to prove a point.

The virus in question was Junín, an arenavirus named for the town in Argentina where it was first described. It causes a severe and often fatal febrile illness called Argentine hemorrhagic fever (AHF). Like other arenaviruses, such as Bolivian hemorrhagic fever, Junín virus chronically infects rodents, and as these rodents come in contact with man, they infect humans. The rodent reservoir of Junín virus is a mouse-sized animal that lives in the grasslands of the fertile pampas of Argentina. This is the same agricultural region that put Argentina at number thirteen in the world in gross national product around the time of World War II, before political upheaval cut down the economy. The parallel in this country would be the plains of Kansas, where there are crops in the fields but no trees as far as the eye can see.

This is important because, unlike in the Beni of Bolivia, where they could simply monitor rodent populations and trap them to keep numbers down in the towns to control disease, the people of Argentina could not avoid rodents during harvest time in the pampas. Nor could they stay out of the pampas, the economic heart of the country.

As with many diseases, AHF was first discovered by a local clinician. In the town of Junín in the early 1950s, Dr. Aribalzaga described a disease neither he nor anyone else had ever seen before. He characterized it as a new epidemic disease with symptoms including fever and hemorrhage. The disease was occurring at the rate of several hundred cases a year and its mortality was as high as 30 percent.

At the onset of the disease, patients experienced fever and body aches; they gradually felt worse and worse until after about four or five days they found they could hardly keep moving and would go to the doctor. In addition to the fever, they had flushing of the face, chest, and back that looked like sunburn, an indication that whatever they had was affecting the small blood vessels in the skin all over their body. Their blood pressure was a little lower than normal, possibly leading to postural hypotension (if they stood up too quickly their blood pressure would dip and they'd feel dizzy). The blood vessels in their eyes would be dilated.

In the next day or two they began to develop subtle signs of bleeding: tiny hemorrhages could be seen in the skin (petechiae) if you looked closely. The nervous system was affected next. Patients would be frightened and begin to lose their orientation to time and place;

they had tremors in their hands and mouth. Their blood pressure would drop further, and the bad cases would fall into shock and hemorrhage from multiple sites around the body, including the gums, the uterus, and the urinary and gastrointestinal tracts. They would begin to vomit blood and develop large purplish blood spots beneath the skin called ecchymoses. Many went into convulsions and even coma. Then they died.

It was a scary disease, to say the least. Most hemorrhagic fevers have an acute onset, but arenavirus hemorrhagic fevers tend to develop slowly, which makes taking a good patient history critical since often patients report they've been sick only a day or so. Yet you would discover their appetites had trailed off several days earlier and they really hadn't felt well for a week.

The incubation period was about ten days. From around day ten up to day fourteen of the disease, patients either died or got better.

Of those who survived, most were believed to have recovered completely. Sometimes a patient who experienced hemorrhage in the brain during the acute phase of the illness suffered residual brain damage, but that was rare. It was more common that a survivor's hair would fall out, almost always growing back later. Neurological problems generally disappeared over time, but survivors often reported dizziness, or a vague sense that they weren't as sure-footed as they were before their illness.

After its discovery, AHF outbreaks occurred every year, involving hundreds of cases, and the endemic zone was expanding, putting more and more people at risk. The disease became a cause célèbre in Argentina as scientists rushed to find a way to protect their countrymen.

Two groups tackled the disease competitively: one from the University of Buenos Aires and another, also in Buenos Aires, at the Malbran Institute. The rivalry between the university and the institute was strong. It once actually erupted into a fistfight on the floor of the National Academy of Sciences in Buenos Aires.

Scientists at both the university and the institute isolated the virus, but officials at the university lab charged that those at the Malbran hadn't been successful. To defend the honor of his institution and prove they had a valid isolate, the brash young scientist Julio Barrera Oro injected himself with the virus. He threw himself a party, where he announced what he'd done, and before he was even fully over his hangover he started feeling symptoms of AHF.

Barrera Oro was an incredibly focused, single-minded individual.

Although some view his actions as virtually suicidal, I think it would be more accurate to describe his motivation as pride mixed in with good, old-fashioned Latin machismo. This disease represented such a scourge to his people that in the worst-case scenario (if he died), he would become a martyr for all time in Argentina. I don't think his cause suffered any when after a period of very severe illness, he did manage to survive.

The virus continued ravaging Argentina, and in the 1960s and 1970s annual figures of up to 1,000 cases were not uncommon, with a mortality rate ranging from 5 to 30 percent.

In 1976 an international conference was held in Buenos Aires, entitled "The First International Seminar on Hemorrhagic Fevers Produced by Arenaviruses." It was clear to all that the only way to control this disease would be through development of a vaccine.

No one had ever developed a successful, officially sanctioned vaccine against an arenavirus before, which was a formidable obstacle. Each virus is different, but once you've found an approach to one virus in a family you have made significant progress in developing a vaccine for the other members of that family as well. To oversimplify, it's like finding a recipe or a road map that you can apply to others in the specific family of viruses. Scientists were ignorant of the basic mechanisms of immunity with these viruses. At the conference, though, it was decided this would be an international priority.

There had been an earlier, controversial attempt to develop a vaccine, led by a scientist from the University of Buenos Aires, Dr. A. S. Parodi, who headed one of the groups that isolated the virus and developed what he considered to be a vaccine against AHF in 1967. His vaccine work was not recognized by the scientific community because of irregularities in his approach, long-term safety concerns, and the fact that it was not a repeatable experiment. Once his "vaccine" was gone, there was no way to produce more.

Parodi had collaborated with prominent American arbovirologist Jordi Casals at Yale to culture several generations of the virus (called passages) in mice. Guinea pigs inoculated with the passaged virus did not develop AHF and were protected when challenged with the original isolate, leading Parodi to believe they had a less pathogenic variant of the virus. He made several clones to serve as a live-attenuated vaccine and took the third clone with him back to Argentina, where he made a large pool of the virus in suckling mouse brain.

This is an absolutely terrible idea for making a vaccine, no matter what family of virus you're dealing with. First of all, there's no

master seed you can go back to if you want to produce additional
lots of vaccine later. Second, this clone 3 virus, as it came to be
called, produced in mouse brain, could harbor any number of con-
taminating mouse viruses. Worse, since it was made in the brains of
these mice, recipients also got a dose of foreign brain material, with
the risk of autoimmune encephalitis. This was a pretty scary premise
for a modern vaccine.

But from Parodi's point of view, something was better than nothing. People were dying in his lab and in the pampas. In fact, on the
other side of town at the rival Malbran Institute, they have a brass
plaque on the wall commemorating the virologists who've died
working with Junín. This was not a theoretical problem to Parodi.
He was looking head-on at the question of how many soldiers you
have to sacrifice to win the war.

Parodi inoculated seven people in his lab and they all developed
antibodies and had no untoward effects.

His next move is where I really begin to split with Parodi on the
bases of both science and medical ethics. He went out into the endemic zone and in 1968–70, inoculated more than 600 people with
his new "vaccine."

I know people were desperate for a way to combat this disease, but
by no means was this vaccine ready for human trial. Laboratory
workers and virologists who volunteered to be inoculated were a
different story: they knew enough of the risks to give true informed
consent. People in the countryside did not have the background to
know what they were getting into. And ethics aside, it was sketchy
science because it was not repeatable.

Still, on one level the experiment seemed to work. No one got AHF
and the people developed antibodies in an immune response that
suggested they were protected. No one died from any side effects and
no one in that group died from AHF after working intensively with
wild-type virus. Thus, it most likely did save lives.

In studies of 213 of the 629 people who received the inoculation, it
was found that a little more than three-quarters developed a reaction
of some type, typically including a fever, often along with asthenia,
myalgia, headache, and retro-ocular pain. In most cases, the symptoms appeared three to ten days after inoculation and cleared up
within another few days. One hundred and sixty-five people were
tested one to three months later and 97 percent had neutralizing
antibodies against Junín virus. Seven to nine years later, follow-up
studies were conducted on 267 and it was found that no one suffered

any long-term side effects and 90 percent still had measurable antibodies.

But it could not be produced again and, even if it could, it was not a safe way to make vaccine. There was probably a lot more luck than science behind the fact that no one got sick or had long-term side effects. Since Parodi, however, there have been no more scientific gains made toward a vaccine against the disease.

When the scientific community decided to work together to make a true vaccine in 1976, Gerry Eddy very much wanted to be a part of it. He was not able to commit USAMRIID's resources himself, but he persuaded Barquist that we should participate.

There were several reasons why it would be valuable for the army to help develop this vaccine. The Pan American Health Organization was going to serve as an umbrella for the group, funding at least part of the research. Working in conjunction with them and other such organizations to develop a lifesaving vaccine would help dispel remaining concerns over "Fort Doom" and give USAMRIID scientific respect and recognition internationally. There was also much to be learned from developing an arenavirus vaccine.

There were strategic military reasons to get involved too. It had been shown that the clone 3 vaccine also protected monkeys and guinea pigs against Machupo virus. This could give us two vaccines in one, providing protection against one of the key biological warfare agents studied at the time.

Barquist approved the plan to bring in an Argentinian scientist to take advantage of the facilities at USAMRIID to develop a vaccine. Given the political infighting among Argentinians, we knew a vaccine completely developed by outsiders would stand little chance of acceptance. We would need to test this in Argentina someday. If we expected any cooperation, this couldn't be a "gringo vaccine."

We realized we needed a special suite dedicated to the effort. You can't make vaccine in a lab full of virus, and you need specially filtered incoming air.

There was only one room that would work. Coincidentally, it was a room Gerry had long regarded as underutilized. It was then in Ken Hedlund's possession, the bacteriology division's; and Gerry had previously attempted to get it out from under him. This time he was successful. The fact is, Gerry was one of the best political chess players around. He managed to position USAMRIID in a joint effort with the Argentinian Ministry of Health and Social Action and the Pan American Health Organization in a United Nations Development Program and, in the process, had also brought more attention, resources, and prestige to his division at USAMRIID.

The next step was to choose the Argentinian scientist to come on board at USAMRIID. The intellectual war was still on between the university and the institute in Buenos Aires and there was a lot of jockeying around. The primary candidate from the university was a talented young virologist named Mercedes Weissenbacher. Although only in her late thirties, she already had a solid reputation. The other primary candidate was Julio Barrera Oro, the mad scientist from the rival Malbran Institute who'd survived his Junín infection and had a very personal reason to beat this virus. In the end, Julio got the position, in part because Mercedes seemed to back away from spending two years or more in the States.

Julio is, to put it mildly, quite a character: enthusiastic and dogged when it comes to solving a problem, with a terrific sense of humor, which is critical in a field as full of frustrations as ours.

There's no specific training or degree in vaccine making, but some people seem to have more of a knack for it than others. Experience helps a great deal because there are approaches and ideas that aren't in any manual that people who've done it successfully think of—like a master chef knows how to mix in a pinch of this or a spoonful of that. Julio had worked with one of the great names in polio vaccines, Joe Melnick, and had absorbed a lot of the lore plus what the USAMRIID crew contributed. It also takes a certain kind of personality: a creative, driven problem solver; someone who can rise out of the ashes of a failed attempt without being too frustrated to see where to go next. Even more than in other areas of research, you don't often get instant gratification.

We decided the new vaccine should not use clone 3 because of its passage history: Argentinian mice were judged unsuitable for producing a safe vaccine for human use because they are not kept under controlled conditions and could carry adventitious viruses. It was also important that any live-attenuated vaccine come from a well-identified strain that could later be used as the parent strain for future vaccine work.

Beginning with the original virus isolate from Parodi and passaged in Jordi Casals's lab (known as XJ-43, the forty-third passage of the XJ strain of Junín), Julio manipulated the virus through another passage (or generation). He single-handedly did additional passages in cell culture, and some cloning, and came up with three candidates for the vaccine.

Julio was like the proverbial kid turned loose in a candy store. He was experienced in classically prepared vaccines from a fellowship working with polio in Joe Melnick's lab at Baylor, and at USAMRIID he had all the latest technology at his disposal. His insight and experience were critical. For all the developments we've made in molecular biology and biotechnology, with only rare exceptions all the vaccines we receive today are classically prepared vaccines.

At every step in development, Julio rechecked his work, making sure all the cultures were tested and retested to rule out the presence of other agents that could contaminate the vaccine virus. He was well aware of the kind of documentation and standards required by the FDA for approval.

Before we could think about testing in humans, however, the vaccine had to be tested for neurovirulence—that is, to determine if it could damage somebody's brain. This was a real concern given the neurological symptoms people got from AHF—the tremors, coma, and convulsions. In the standard test for neurovirulence the vaccine is injected into the brain of a monkey to see if it induced lesions. But first you start with smaller animals. We found that the vaccine didn't cause any neurovirulence in guinea pigs, even baby guinea pigs infected directly into the brain. We got the same result from rhesus monkeys, so we were beginning to believe the vaccine was safe for use in humans.

The next concern was testing for chronic or latent infection. This is especially important with arenavirus vaccines since these viruses always cause chronic infection in their reservoirs and often in other animals as well. Once again, the vaccine passed with flying colors.

We also checked humans for latent or chronic infection with the unmodified virus. Another well-known, well-respected Argentinian scientist named Julio—Julio Maiztegui—was a close collaborator in the field. There were enough people in the endemic region who'd survived AHF so that occasionally one would come to autopsy after some kind of accident. One time, for example, Julio M. acquired tissue from a fellow who'd had AHF years earlier and then had to have his spleen removed for a completely unrelated reason. Julio tested the spleen, and also samples from another guy who'd been run over by an auto. He found no evidence of latency. So we were increasingly confident that even the original virus didn't have long-term persistence in humans.

For scientific as well as political reasons, every experiment or test done with the vaccine Julio was developing was also done with clone 3, so we could show that whatever we were making was better than the vaccine that already had been safely used in 629 people in Argentina.

As talented as Julio was, all was not smooth sailing. At one point, he cultivated spleen cells from some guinea pigs he was testing and got an agent that looked like AHF. Was this evidence of chronic infection in guinea pigs after all? Fortunately, when the agent was examined under the electron microscope, it was found to be a latent guinea pig herpes virus, completely unrelated to the vaccine.

One of the big selling points of the project had been the idea that we could get two vaccines developed for the price of one. So at the same time as the AHF work was going on, we were conducting studies to see if this vaccine would protect against Machupo, and it did.

Since the army's main concern was developing something to protect against airborne virus, and our initial tests on guinea pigs were done by injection, a specialized group was brought in to expose the animals that way. These experts not only knew how to prepare the aerosols for introduction but could also calculate and characterize different doses of aerosols.

When they exposed the vaccinated animals to aerosols, all the guinea pigs were dead in twenty-four to forty-eight hours. After we tied Julio to a chair (he could be a bit volatile sometimes), we discussed what could have happened. He was concerned that somehow the vaccine had set the animals up for a pathological reaction to the wild-type virus that was killing them all.

Up to this point, we'd all contributed ideas as a group of about a dozen people, including Gerry Eddy (before he retired partway through), another virologist named Kelly McKee, and me, in planning the experiments. But Julio was the main developer of the vaccine. Now, however, all his test animals were dead and he needed a fresh point of view to get back on track.

I asked him what, exactly, the guinea pigs were inoculated with, and what had been in the virus used to challenge them. It turned out the different doses of both inoculated virus and aerosolized virus had been diluted in calf serum, a common practice. Now, I was not entirely unfamiliar with calf serum—it was, after all, a driving force behind my decision to work at USAMRIID in the first place. I wondered whether the guinea pigs were having some kind of allergic reaction to the serum unrelated to the presence of virulent Junín virus. If so, they would have developed a sensitivity to calf serum from the inoculation, priming them for a hypersensitivity response when they got the aerosol.

We tried the test again without the calf serum, and this time the vaccinated guinea pigs were completely protected, while the unvaccinated controls died. We repeated the test in monkeys and the results held.

When we were finished, we submitted a three-foot-thick sheaf of papers to the FDA and got approval to try this vaccine experimentally in humans. Our plan was to announce our progress at a big virology conference in Argentina in 1982.

Julio's presentation was a joy to watch, and he was received by former colleagues and countrymen with much hugging and applause. We all shared in the celebration onstage. I thought we'd scored a major public relations victory for the new vaccine.

What I didn't foresee was that, concurrent with the meeting, a scandal was unfolding regarding the testing of another vaccine, unrelated to our work.

The Pan American Health Organization—the same group we were counting on to bring international respect and acceptance to our project—had a zoonosis center in Buenos Aires run by a D.V.M./ Ph.D. named Joe Held, who had been Assistant Surgeon General in the Public Health Service before he retired to run the center. One of his responsibilities was to test animal vaccines. Rabies vaccines were vital because the disease is the source of tremendous economic loss in Central and South America. Vampire rabies, spread by chronically infected bats, causes widespread cattle illness and death throughout the region.

Joe visited the Wistar Institute in the United States, where they were promoting a new vaccine against rabies that was truly cutting-edge. The vaccine was set up as sort of a piggyback of viruses. You inoculated an animal with one virus that carried a gene from another virus. The idea is that the first mild attenuated virus will infect the host and then, in the process of infection, it will reveal the gene from the other virus and trigger an immune response to that virus as well. Essentially, this approach would allow you to create the antibodies for a virus without ever having been exposed to the disease itself. It was known as a vectored vaccine.

In 1982 this was brand-new, and when Joe got the chance to test this, he jumped at it. He took it back to Argentina and wrote up some protocols to get a trial started. Now, in the United States you would need review and permission by a dozen different agencies to test this outside a laboratory setting. In fact, they were working this stuff in BSL 3 facilities at the Wistar. None of these agencies exist in Argentina, however. Joe evaluated his protocols and felt he'd worked up a safe experiment well within the brief of the zoonosis center. Unfortunately, it went awry.

The way things were supposed to work, people would be hired by the zoonosis center to care for cattle corralled in an area outside Buenos Aires called Azul. The workers would be vaccinated against both vaccinia virus and rabies, so there was no question of any risk. The cattle would receive the vaccinia vectored rabies vaccine, and samples would be taken locally and sent to the zoonosis center so their antibody response could be measured.

Several things went wrong. First of all, the workers in Azul never got vaccinated. Second, they reputedly started doing things not specified in the protocol, such as milking the cows and selling the milk in town. When this rumor circulated, the already fractious and politicized scientific community went ballistic. Joe was pilloried in editorials everywhere.

One of the presenters at our meeting turned out to be a prominent vaccinologist at the Wistar Institute named Stanley Plotkin. He'd not been warned of the scandal and the place rapidly turned into a shooting gallery. He was at a complete disadvantage because he hadn't heard the issues and because, although he'd once lived in Argentina, his Spanish was a little rusty and he couldn't keep up with the debate (at one point, confusing the word for skin with hair, he began talking about vaccinating in the hair). One particularly wild and ridiculous gentleman asked, "Do you know what's been done to us in our country, violating our national sovereignty with this vaccine? It's worse than detonating an atomic weapon because with a weapon there's fallout and it's over. But here you've released a living organism in the environment and it will never go away. It'll spread and take over the entire country!"

One of the responsible Argentinians finally stopped the proceedings and recommended an hour's break, after which time they'd arrange a special roundtable session for Dr. Plotkin to answer all questions. Poor Plotkin was lucky to get off the stage, sweating and flushed, but I knew we had to act fast if we were going to make any inroads at salvaging the situation for the afternoon. I was a man driven—not just by altruistic pity for the sacrificial scientist but also because I could see the political pendulum swinging against us. Our new vaccine could rapidly become a casualty of the entire affair.

We gave Plotkin a quick lesson in the rabies scandal. It helped him to know why he was being attacked, but he wasn't really able to gain any ground in the afternoon and it started to look like our carefully developed, diplomatically positioned vaccine was in deep trouble.

The situation got even worse as the result of a Soviet-spread disinformation campaign. A scientific hack announced to a British newspaper that the U.S. Army (specifically USAMRIID) had synthesized and manufactured the AIDS virus and turned it loose in Africa for testing. How did the people of Argentina know our vaccine wasn't just another HIV experiment?

It's difficult to counter an accusation like that, particularly if you work at Fort Doom. There are two schools of thought in the government on managing accusations like this. The army's general attitude is that we don't talk to the press, we issue official statements. Everything has to be cleared all the way up to the top, which means an awful lot goes unsaid. CDC has a very different, much more proactive attitude. They get out and work the press.

I didn't have a press officer, and there was no one from the army to advise me. If there ever was a time in my stint there where I asked myself, "Do they pay me for this?!" this was that time.

I was saved strictly by luck. The first publication to carry the accusations was a well-known British newspaper, which was picked up in the Argentinian press. While sitting in a café on a quick lunch break, I was spotted by a local reporter with a TV cameraman in tow. Short, slightly chubby, with an ascot and very well-oiled, slicked-back hair, he represented tremendous trouble for all of us if I handled him the wrong way.

He started asking me about the accusations in Spanish, and I figured I had three options: I could refer him to the American embassy, I could try to deal with everything myself, or I could try to deal with part of it and refer him to the embassy for the rest. I figured he would be more positively disposed to what I had to say if I tried to be responsive rather than taking the safe route and sending him to the embassy for official comment. So I said in my best Spanish, "Look, I work at USAMRIID and I'm a virologist. I know what we can and can't do and we are not smart enough to make HIV from nothing. HIV is a very complex virus that we don't know how to deal with. We don't know how to vaccinate against it, we don't know how to cure it. Our country is in the throes of a tremendous problem with

HIV. It's even in our blood supply and we're trying to deal with that. This is not something we could make or did make."

The reporter broke into a smile. "That is what I wanted to hear you say. I did not believe these charges. I know the paper that carried this in Britain. They were against Argentina during the Falklands War. They called us 'Argies' and made all sorts of insulting remarks. I didn't believe what they said then and I don't believe this at all." Eventually the whole thing blew over, but it confirmed for me that if we hadn't had Argentinian scientists involved from the very beginning and if we hadn't had an international organization that had some credibility involved all along the way, we would have been dead in the water.

By the time we were ready to leave the conference, the three junior scientists I'd taken with me, who'd played important roles in the vaccine testing and development and who'd been so anxious to go they'd been hounding me for months with offers to carry my bags, were completely disillusioned. One of them earnestly turned to me and said, "You can have all that stuff. I'm gonna stay home and work in the lab. This political stuff is just too hard."

When we got back we were ready to test the vaccine in humans. We were very sensitive to the lessons we'd learned at the conference and figured the safest approach, politically, would be to test the vaccine first on U.S. volunteers in experiments run by an Argentinian clinician and an Argentinian scientist. We asked Julio Maiztegui to recommend people to us, and when the time came for the first tests, we had an Argentinian clinician side by side with us on the protocol, inoculating four gringo volunteers with a vaccine developed by a famous Argentinian virologist. No one was going to accuse us of experimenting on the Argentinian people.

It is a tremendous risk when you make the leap to experimenting in humans, especially when you're dealing with a live-attenuated vaccine. We'd tested and retested the vaccine every way we knew how, but the fact is you just never know what will happen when you inject this stuff into people. If you think it's hard looking into the eyes of a monkey you're going to inject with a new vaccine, try doing it with someone's twenty-year-old son.

As so often happens in science, we ended up with mixed results. The vaccine proved safe, which was obviously good, but the volunteers had a poor immune response. We tried it on more volunteers and got a similarly weak antibody response using our standard tests. Julio was considering jumping off a cliff until we tested all the volunteers for T-cell immunity and found the T cells were much more reactive than we expected. Julio meanwhile developed a more sensitive test for antibodies and this time it looked good. There was more responsiveness from the immune system than we could see with less sensitive tests.

Medical research can be unnerving because as soon as you get a positive outcome from one experiment or test, before you can rejoice or even breathe a sigh of relief that it's working, it's time to up the ante. You can't test a new vaccine on ten people, get good results, and judge it safe for four million Argentinians or the 82nd Airborne. We felt that for this vaccine ever to be accepted by the U.S. and the Argentinian scientific and lay community, we'd have to do a field trial to prove it worked. The trial had to be done randomized, double-blinded, and placebo-controlled. We did what we call a power calculation and figured that for a reasonable chance of showing reasonable efficacy, we needed at least 6,000 people in the trial: 3,000 would get the vaccine, 3,000 a placebo.

We were also concerned about late side effects. The animal data was okay, but we weren't willing to put this vaccine in 3,000 people and then watch them develop some unexpected late complication. We wanted to rule out large, serious reactions first and calculated we'd need to test a couple of hundred people to rule this out at even the 2 percent level. With a lot of help and preparation (and no small amount of PR) Julio Maiztegui gathered a couple of hundred Argentinians to begin the second-stage trial while we continued to inoculate North American volunteers.

As another safety precaution, we restricted the trial to the people at greatest risk of exposure to the virus. The high-risk group was made up of adult males working in a particularly "hot" area of Santa Fe province. These people had the most to gain if the vaccine worked, so they bore the greatest risk up front. The trial was again a success, with no significant adverse reactions in more than 95 percent of the recipients. Parenthetically, one person in the trial was later found to have died from HPS, one of the first indications this disease would be a major health problem in South America.

The army provided a large amount of vaccine to get the Argentinians started. Since then, the vaccine has been given to about 125,000 people and has continued to be targeted to the populations at highest risk. The Argentinians are now trying to put together a vaccine production facility that will allow them to manufacture vaccine themselves. In the high-risk populations, the number of cases of AHF has fallen to less than fifty per year, as opposed to the several hundred per year prior to the vaccine. And there have been no significant complications from the vaccine. We hope we have given the Argentinians the means to control the disease and that, as vaccine coverage increases, the number of cases of AHF will continue to decrease.

We still have not done studies in humans that would show whether the vaccine protects against Machupo, but it has been very successful in preventing BHF in animals. After the AHF experience, we no longer had personnel or funding to do human studies, and they should really be done in Bolivia, following the same approach of vaccinating the people at greatest risk of contracting Bolivian hemorrhagic fever.

In addition to proving the vaccine's effectiveness and safety in humans, and its potential for protection against Machupo, the project met another one of our goals. We now have at least one approach for developing a vaccine for an arenavirus. It may not work for the next arenavirus we confront, but at least we have a proven option where before we had none.

This was my initial hands-on experience working to develop a vaccine from scratch. I also used my time at USAMRIID to learn about how a much larger organism worked: the U.S. military. Even within the safe confines of USAMRIID, I learned how to maintain my autonomy and get things accomplished within the bureaucracy.

I saw that rank was earned in one of two ways: doing a good job and getting recognized for it, and politics. Some of politics is attention to detail: making your boss look good, picking the right assignments. Some of it is kissing the right butts and screwing your competitors. I like to think I got my promotions by earning them, and that the little politics I'm sure I've engaged in were of the former and not the latter.

It's very different from bureaucracies I've encountered in the Third World, where you find you can sometimes be more successful by doing nothing than by going out on a limb and risking making a mistake. If you make your boss look good in a bureaucracy, for the most part you'll both look good. If you make him look bad or threaten him, the relationship (and often your next promotion) is doomed. In both environments, though, the painful lesson I learned from John Seal at NIH holds true: no matter how well you do your job or what your intentions are, if you make your boss or your organization look bad, you're in trouble. It can also be hard to get ahead if you want your boss's job. I've never lusted after any of my bosses' jobs and I think they could sense that.

Some of the lessons I learned there have only become visible to me as they've been put in context by subsequent experiences at CDC. For example, I know now that even within the caste system, different departments and compartments in a bureaucracy hold more power (read: budget, in many instances) than others. At USAMRIID, Gerry Eddy positioned virology well to grow rapidly by bringing in talented people. He was fortunate in that his boss, Dick Barquist, shared an interest in seeing that division grow. As I've seen firsthand how difficult it is to recruit and build in times of shrinking budgets and shifting political agendas, I have to respect what he accomplished in a very short period of time.

I've been very lucky because in most of the places where I've worked the price for doing science has not been too high, and even with all the political pressures at work, I've been able to keep a reasonable degree of scientific freedom at a time when it's been very rewarding to be a scientist. When I look at the technology we have at our disposal today, the tools for communication and research, I wonder what true geniuses like Snow, Reed, and Finlay would have been able to accomplish. Rather than being diminished by the relatively simplistic and crude conditions under which they worked, their accomplishments are only all the more incredible and impressive to me.

It's easier to reach for the stars when you stand on the shoulders of giants.

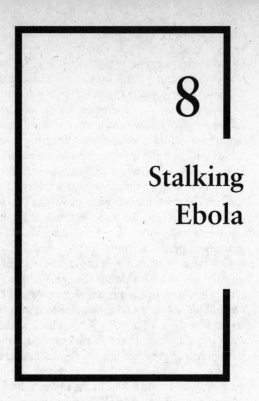

8

Stalking
Ebola

THE REAL FEAR, of course, is that some mysterious infectious disease from a remote and primitive jungle will break out and run rampant in an industrialized city in the Western world, despite all the sophistication and technological acumen of its inhabitants. After Michael Crichton's best-seller *The Andromeda Strain* was published, Karl Johnson (and, perhaps because of his influence, others in the field) began using that term as shorthand for a major outbreak of unknown origin that would, at least at first, overwhelm our powers to halt it.

Ironically, the first modern Andromeda Strain type of scare occurred in August 1967, a full two years before Dr. Crichton's novel was published. And unlike the fictional bug, which came from outer space and visited itself upon a small town in Nevada, the real one came from somewhere in equatorial Africa and afflicted an industrial city in the central region of what was then West Germany. The city was Marburg, an ancient Hessian university town on the banks of the Lahn, with a population of 75,000, graced by a magnificent Gothic cathedral and a fourteenth-century castle, about fifty miles due north of Frankfurt. The disease that struck there went by no name at all, because no one had ever seen it before.

Generally speaking, outbreaks and epidemics are apparent only after they are under way. There is usually one dedicated, methodical, and observant physician who first recognizes a pattern and, like a good detective, begins putting together the clues in his or her own mind. In the 1994 Four Corners outbreak, it was Dr. Bruce Tempest of the Indian Health Service. And in Germany in 1967, it was Dr. Gustav Martini, professor of medicine at Marburg University.

Dr. Martini noticed first one, then two, then three patients at the University Medicine Clinic presenting with "flu-like" symptoms. But it didn't stop there. They went on to develop agonizing headaches, high fevers, and worsening pains all over their bodies. Martini was the first one to make a connection between these patients, to try to figure out what they had in common. This wasn't influenza, he quickly concluded. This was something he'd never seen before. And the patients turned out to have a common exposure that was decisive. They were all employees of Behringwerke AG, a pharmaceutical company that produced vaccines. These three patients had all had contact with African green monkeys, also known as vervets, or in scientific nomenclature *Cercopithecus aethiops,* which had been imported from Uganda for use in medical research. Monkey kidney cells were used in the production of polio vaccine, one of Behringwerke's main products.

In the course of the following three weeks, seventeen more patients were admitted to the Marburg University clinic. And two women—a doctor and a nurse—became infected inside the hospital itself. Similar outbreaks occurred in Frankfurt and in Belgrade, Yugoslavia, all apparently related to the handling of monkeys shipped from Uganda. The sequence of symptoms went on to include bloodshot eyes, dizziness, nausea, pain in the chest and abdomen, and severe vomiting and diarrhea. Then, as the days progressed, it grew worse still, with spreading subcutaneous hemorrhages, blood seeping from various body orifices, and, in many cases, destruction of most of the liver. Patients were going into shock and bleeding to death. The kill rate was about one in four—a very scary disease.

As soon as he began to see the epidemiological pattern, Martini got in touch with Behringwerke and had them begin sending blood and tissue samples to the diagnostic lab at the university. And there, Werner Slenczka and his colleagues, working under lab director Rudolf Siegert, accomplished an admirable feat of medical virology.

Without sophisticated biosafety facilities, using masks, gowns, and gloves and working on tabletops without laminar flow hoods or pressurized-air suits, they carefully adapted classical procedures which, with refinements, are still the model for the identification of unknown biological killers.

The first step was to inoculate cell cultures with samples from patients and allow the virus to spread throughout the layer of cells until they were expected to be infected and contain antigens of the unknown agent. As soon as they stained one of the slides with routine dyes, they saw under an ordinary microscope some strange, suspicious-looking inclusions in the cells that appeared to be products of viral replication—specifically, reddish-blue dots that looked like inclusions other scientists had described in other virus-infected cells. At this point, they had no idea what the cells were infected with, only that it was causing severe disease or death. Next, they drew blood from patients who had gotten sick but survived the infection, since, presumably, they would have developed antibodies against it. That blood was spun down in a centrifuge to separate serum, called convalescent serum or, simply, antiserum. The antiserum was then added to the tissue slide and later thoroughly washed. Through this process, they washed away anything that didn't bind tightly to the antigen, such as a specific antibody. They then added antibodies tagged with a fluorescent dye to detect whether the antiserum had bound to the antigen.

The next step was to take the slides to a special microscope that would shine an ultraviolet light on it and make the fluorescent dye visible. They darkened the room lights, and, to their enormous satisfaction, they saw a visual link between the sick people who had recovered from the disease and the agent that was growing in their cell cultures.

When they saw the results, they thought they had something significant. But before they could be sure of the meaning of the bright fluorescence, they had to compare it with a control. So they looked at a slide of normal cells treated with the convalescent serum under the same scope and saw only the faintest of shadows. Similarly, they applied normal serum to the infected cells and this, too, came up negative.

So far, so good. Then they turned to the electron microscope to see, under its enormous magnification, exactly what it was that was growing in their cell culture. Electron microscopy involves both science and art. The procedures are extremely complex and exacting and any small vibration or electromagnetic interference can render the results useless. Therefore, most of these units are placed in basements, directly on thick, completely stable floors and surrounded by overlapping layers of copper mesh all around the room, known as a Faraday cage. They subjected another infected cell culture to the complicated procedures needed to clearly visualize the tiniest structures that can be found within a single cell. When the Marburg researchers sat in the semidarkened electron microscope room and stared at the green glow of the screen, they saw the image of something that no one anywhere had ever seen before.

The virions—the individual particles—of most viruses are more or less spherical, looking like tiny tennis balls, or peppercorns, or, in the case of arenaviruses, spheres containing grains of sand. But here they were seeing virions that collectively looked like tangles of string or spaghetti and individually resembled tiny shepherd's crooks. These German scientists were looking at a new virus; no one had ever seen a virus that looked like this before. And in the ensuing weeks, months, and years, the virus whose individual units they were seeing here became known as vervet monkey, green monkey, or, finally, Marburg virus, following the custom of naming after the area where it was isolated.

Impressively, these results and the identification of this new and dangerous virus were achieved before the final surviving patients from the outbreak had left the hospital.

The outbreaks in Marburg, Frankfurt, and Belgrade died out on their own. And though scientists now knew they were caused by a virus and what the virus looked like, in some ways the mystery was as tantalizing as it had been when Dr. Martini first noticed something odd in the university clinic. From a public health standpoint, the end of new cases was obviously very good news, because people were no longer getting sick and dying from Marburg disease. From a science and medical research standpoint, however, it was both troubling and perplexing, because this also meant that nobody knew where the disease was "hiding." African and Asian monkeys are critical to laboratory science throughout the Western world; you couldn't simply stop importing them. Even worse, cell cultures from their kidneys were used to prepare polio vaccines and the polio virus could grow side by side with Marburg in the same cell cultures. Here was a disease which had proven its ability to jump species, very bad news in and of itself. And if, as was estimated, the European outbreaks stemmed from perhaps as few as two or three infected monkeys out of a shipment of more than 500, then there was an unsolved problem which could become critical again at any time. Whenever something new and deadly comes along, the worry is that this is only the beginning. With the mysterious Marburg monkey disease, that turned out to be tragically true.

There was no more to the story until early February 1975, when two young Australians, a nineteen-year-old woman and a twenty-year-old man, were hitchhiking around southern Africa. In Rhodesia, now known as Zimbabwe, the man felt pain and noticed swelling in his right flank and figured he'd been bitten by some kind of insect. But six days later, on a beach in South Africa, he broke out into a severe sweat and suddenly felt extremely fatigued. Before long he was also feeling nausea and dizziness, pain in his other leg and shoulders, and had difficulty breathing. On February 15, he was taken to a hospital in Johannesburg with a provisional diagnosis of typhoid fever. There, a large team of doctors, nurses, and laboratory researchers studied his illness and gave him aggressive, first-rate medical support, but his condition worsened and he did not respond to drugs. Four days after his admission to the hospital, he died from shock due to massive internal hemorrhaging.

Two days later, his girlfriend also came down with similar symptoms, and a week after that, one of the nurses who had been treating him also became sick. This clearly was not typhoid, but it looked as if it could be Lassa fever, so both women were treated with Lassa antiserum. They were also treated with the anticoagulant drug heparin.

It might seem counterintuitive that an anticoagulant would be effective as a treatment for widespread internal bleeding, but it goes directly to the core of the problem in treating some hemorrhagic fever cases. If you cut your finger, the platelets in your blood will start the clotting process so that the bleeding stops. Of course, the clotting has to be localized to the site of the cut, or else all of your blood would clot, you'd have no circulation, and tissues all over your body would start to die off for lack of oxygen. So the body cleverly maintains a counterregulatory mechanism to the platelet effect which keeps it from going beyond the local action that's needed. In other words, there is a balance between clotting and anticlotting.

What happens in some of the severe hemorrhagic diseases such as Marburg, Rift Valley fever, or Crimean Congo hemorrhagic fever is that many, many tiny clots begin forming all over the body, a condition known as disseminated intravascular coagulation. And everywhere one of these tiny clots forms, a piece of "downstream" tissue dies for lack of oxygen. Trying to balance out this diffuse clotting problem, the body begins sending out the specific activated molecules (plasmin) responsible for taking up clots. The consumption of the body's clotting proteins and actions of plasmin result in defective clotting. We end up with massive bleeding, or blood leakage, throughout the vascular system. So what the heparin is intended to do is to prevent the initial clotting. But obviously, heparin can still cause bleeding on its own, so a physician really has to walk the line on this type of patient management, weighing both the advantages and the disadvantages of intervention, as well as deciding when in the course of the disease it makes sense to try something. For example, there is always a tendency to want to administer some sort of convalescent plasma, even if you don't know whether it's effective, and even if you're not sure of the exact virus you're trying to treat. But there is always a great danger in doing more harm than good, particularly with all of the serious and potentially fatal blood-borne diseases circulating today.

By our very nature, doctors always want to do something, if for no other reason than to make ourselves feel less powerless. This is the crisis I went through back in Bolivia as I helplessly watched Donato die.

The two women in the Johannesburg hospital had a very rough time of it; their symptoms take up an entire column in the issue of the *British Medical Journal* that reported on the case. But with extraordinary medical support, both managed to survive. It's also important to note that, unlike the first case, what is known as barrier nursing was stringently practiced: those attending them always wore mask, gown, and gloves. Of all the sophisticated medical techniques in use today, this simple procedure is arguably among the most important.

During the course of the nurse's illness, the lab studies on the index case indicated the presence of the little-known Marburg virus. Doctors had a supply of antiserum shipped in from Behringwerke to see if that would do any good. It was only a theoretical possibility; there was no real-world experience with this treatment potential. But by that time, thankfully, the second woman had begun to show improvement, so it was never administered.

And then Marburg disappeared again. But the tantalizing mystery remained. How did the young Australian contract the disease? Did it have anything to do with the apparent bite on his leg in Rhodesia, or was that just incidental? If it was related, what caused the bite? Was it an arthropod, as seemed likely, or maybe even a rodent, a small animal? It seemed unlikely that anything very large could have bitten him without his noticing, but there was no evidence that the "reservoir" for Marburg was arthropods.

So where was it from, and where was it hiding?

Then came the big one.

Like all of them, it started out small. An old and well-respected Catholic mission hospital run by Belgian nuns of the Sisters of the Sacred Heart of Mary order in Yambuku, in the rain forests of northern Zaire, saw what turned out to be the index case. He was a forty-four-year-old local teacher by the name of Mabalo Lokela, who worked at the school connected to the mission. He had just come back from traveling with six companions throughout northern Zaire and the Central African Republic and he had a high fever. It was August 26, 1976. Coincidentally, this was the same month that veterans attending an American Legion convention at the Bellevue-Stratford Hotel in Philadelphia were getting sick and dying from a mysterious, flu-like respiratory syndrome. It would be five months before Joe McDade at CDC figured out the culprit there was a bacterium that had escaped the scrutiny of medical science for decades.

Leading off the differential diagnosis in Zaire was malaria, a protozoan more complex in structure than either viruses or bacteria. It was a disease Lokela had had before and one which was endemic throughout the region. He was given an injection of quinine. He went home, but since he worked in the mission complex, he returned to the hospital each day.

Two days later, a thirty-year-old man arrived at the hospital with severe diarrhea and nosebleeds. The diagnosis in this case was dysentery, but the man left the hospital after two days, against medical advice, and was never heard from again, representing yet another small medical mystery that was never solved.

Lokela's fever was not getting any better. He'd begun vomiting uncontrollably and experiencing severe diarrhea. He had a headache and dizziness, intermittent confusion, and trouble breathing. Then he began bleeding—from his nose and gums and rectum. The rectal bleeding was arising from diffuse sites throughout his gastrointestinal tract.

The 120-bed Yambuku hospital was a collection of low whitewashed cinder-block buildings with corrugated-tin roofs. Like most outpost hospitals throughout village Africa, Yambuku had no doctors on staff and the nursing staff were not RNs. Still, they were all smart, experienced, and dedicated both to their faith and to their fellow man. They would have to be to willingly put up with the hardships and privations of living in a remote region of a poor and politically unstable country like Zaire. As Lokela's condition worsened, they threw everything they could think of at it—various antibiotics to cover a wide range of bacteria, chloroquine for malaria, hydration, vitamins, anything. Yet nothing helped. He died on September 8, about two weeks after the fever first appeared.

This was a blow to the entire mission community. Mabalo Lokela had been well liked and respected as a man of wisdom and learning. Even so, one isolated, mysterious death in any hospital, particularly in a place as remote as Yambuku, does not generate much alarm. But as was becoming clear, Lokela was not the only one. Others started turning up with similar symptoms, progressing from fever and muscle aches to dire straits in a space of days.

It was customary for the dead to be tended to by close family members. This consideration involved washing the body and cleaning out all food and excreta from the entire GI tract. In Lokela's case, the ritual was performed by his wife, his mother, and his sister. Within days after his funeral, all three women were sick. Even more alarming, nurses at the hospital were coming down ill.

Among the most frightening things that can happen in medicine, one that perverts the recognized order, is when deadly disease breaks out in a hospital, when going to a place of healing causes suffering or death rather than preventing it. Panic spread throughout the surrounding villages. Something at the Yambuku mission hospital was making people there die a horrible death. Eventually, the hospital had to be shut down. When Zairian health authorities visited the place, they saw men, women, and children dying before their eyes.

At just about the same time, word came through to international health authorities that an epidemic with similarly deadly symptoms had broken out in southwestern Sudan. This part of the Sudan shares a border with northeastern Zaire, though the connection between the two outbreaks, if any, was obscure.

Blood and tissue samples from victims of the twin epidemics were sent to the best labs around the world: CDC; the Microbiological Research Establishment at Porton Down, the British counterpart to USAMRIID in Salisbury, England; the Prince Leopold Institute of Tropical Medicine in Belgium; the Pasteur Institute in France. Could it be Lassa? Or a particularly virulent strain of yellow fever? It certainly had some symptoms of both and it was not responding to any of the normal treatments. And the kill rate was terrifying: over half in the Sudan and about 90 percent in Zaire!

The labs ran screening tests. They came back negative for Lassa, negative for yellow fever, negative for typhoid. But when they finally isolated the agent that seemed to be killing all these people, what it looked like was Marburg—that same spaghetti-like, or shepherd's crook or pretzel shape that had been seen in no other virus on earth. If it was Marburg, it had come out of hiding again, this time with absolutely devastating effect.

Karl Johnson was chief of viral Special Pathogens at CDC and he was charged with figuring out exactly what this super-Marburg strain was all about. Karl's team included his wife, Patricia Webb, and Fred Murphy, a D.V.M./Ph.D. who has become one of the world's experts on exotic viruses and the premier creator of electronmicrographic images of those viruses. Patricia worked in the cramped BSL-4 hood line, all that CDC had at the time. Samples were inoculated only into tissue cultures because of the space limitations and safety considerations. Patricia scraped some cells and gave them to Murphy even before any sign of damage to the cells.

What Fred saw under the electron microscope, in a tissue culture prepared from monkey kidney cells, was something that did look like Marburg, but, under Fred's experienced eye, was slightly different in length. They tried to stain it with fluorescent antibodies reacting with Marburg and it was negative. Much later, when genetic sequencing became possible, they found other differences as well. So there was another one out there! How many more? Where do they really come from? This one from Zaire resembled Marburg, but was a whole lot more lethal.

What made it so lethal and how was it spread? Neither was an easy question, but the second one became clear before the first. In an outpost hospital like the one in Yambuku, there is no such thing as a "throwaway" needle of the type we are used to in Western hospitals and doctors' offices. Single-use syringes—single-use anything, for that matter—are a luxury they cannot afford. So the same five or six syringes were used throughout the day on the many patients, with minimal cleaning or sterilization treatments between use. This turned out to be one of the primary methods of transmission. Secondary transmission was achieved by close personal contact, such as between a patient and the personnel caring for him or her, or even more direct contact—the washing and cleaning of an infected dead body. That was one of the reasons so many of the new victims were close relatives, particularly close female relatives, of previous victims.

The deaths were continuing when the World Health Organization released a bulletin on October 15 referring to the Zaire and Sudan outbreaks as "haemorrhagic fever of viral origin. . . . Samples from Sudan and Zaire have revealed the presence of a new virus, morphologically similar to Marburg, but antigenically different."

CDC director David Sencer sent Karl to Zaire to check out the situation for himself. Karl found himself on the plane with Dr. Joel Breman, a young epidemiologist also being sent from CDC, and Bill Close. Close introduced himself when he heard the shop-talk between Karl and Breman. Close was an American physician who was coming to the end of a sixteen-year commitment to Zaire during which he had headed the major hospital in Kinshasa, the national pharmaceutical depot, and personally practiced his own brand of general medicine to the benefit of the people of the country. Karl enlisted Bill as the head of logistics. Close had the ear of Zairian president Mobotu Sese Seko and was able to use his many contacts to pull together supplies and transportation that would be invaluable for the effort. By this time, the international medical community was involved, including the Institute of Tropical Medicine in Antwerp and many others integrated into a WHO team. In Zaire, Karl, Joel Breman, and the others from Atlanta saw what looked like a war zone. The devastation was unreal.

The worst aspect was that there was nothing to do. Other than giving whatever medical support was possible with the modest facilities available, there was no cure and no treatment. The virus antigens didn't cross-react with Marburg antibodies. The best that could be done was to prevent the continuation of the kinds of practices they now knew had spread the disease, and isolate the victims until they either died or got better on their own. And to a far, far greater extent than with Marburg, these people were dying. This was like a wildfire that could not be extinguished; it could only be geographically contained so it couldn't spread and would, hopefully, eventually die out on its own.

That was what finally happened. It could be traced through the epidemic curve. There was an initial rise in the disease as it spread at the hospital. Then hospital personnel started getting sick and later dying. And as people saw that those in the hospital were dying, fewer new cases were going there, so the virus spread decreased. Of course, it was still being spread through person-to-person transmission and the ceremonial handling of the dead.

By the time the international team reached the area, the disease, which had spread to sixteen villages, had virtually disappeared. Hospitals lay abandoned, with many of their staff dead, traditional burial methods abandoned, and a stark kind of quarantine in place—the elders of the villages ensured that the sick entered their homes and did not leave until recovery; the house was burned if they died. The international team drove up and down the old Belgian colonial road looking for new cases to make sure the epidemic was over and conducting retrospective studies.

A few of the cases ended up in the larger cities, such as Kinshasa, but as long as barrier nursing was used in those few individual cases, there was no secondary transmission there.

Altogether, out of the more than three hundred cases reported, nearly nine out of ten victims died. If even one individual in the contagious phase had gotten on an airplane and started spreading the virus in other parts of the world . . . the implications were terrifying to consider.

And at the end of the epidemic, what did the scientific community have to show for itself? For one thing, Marburg was no longer a virus all by itself. This new killer had to have its own name. David Simpson and the crew from England wanted to name the new virus after the village of Yambuku. Karl proposed the name "Ebola," after a tributary of the Congo, feeling that it could represent both the Sudan and the Zaire sites, and was not far from some of the first cases. Stefan Patyn of the Institute of Tropical Medicine in Antwerp, who represented the third group in the neck-and-neck race to isolate the virus, felt CDC had the right to name the virus because they had made the immunological distinction from Marburg first. And so the new virus came to be called Ebola. The virology community also had to come up with a new designation for the virus family, Filovirus, named for the resemblance of the virions to filaments. And they had to accept the idea that if there were three filovirus strains they knew about—Marburg, Ebola Zaire, and Ebola Sudan—there were probably more waiting for the right circumstances to emerge.

I was still at Scripps during the Ebola outbreak. The first I heard about it was at the tropical medicine meeting later that year. They had an audiotape of Karl in Kinshasa, giving a report à la Edward R. Murrow's firsthand account of the Battle of Britain. It was extremely dramatic and everyone in the room was galvanized.

I remember thinking to myself, *Oh shit,* at the effects of this horrific disease. But there was also the sense that "I missed one of the big ones." Nonetheless, I knew, in my heart, that it would be back. It was just a question of time and conditions.

The virus had been isolated, its effects on human populations were now known, and its means of transmission at least partially understood. But the exact mechanism by which the filoviruses caused such severe disease was not known. Nor did we know the reservoir, presumably some forest animal. There was some speculation that it might be a particular species of bat, or perhaps spiders, ticks, or monkeys. It was known that monkeys could transiently carry the virus and transmit it, but did it originate with them or did they get it from somewhere else? When you stopped to think about it, the monkey reservoir didn't make all that much sense, because the disease was so lethal to primates. If it was depending on monkeys as its natural host, we should have seen evidence of monkey disease when we saw human disease. All of this remained a mystery.

A further complication in the puzzle was the lack of a neutralizing antibody response in convalescent sera from human or animal survivors of filovirus infections.

If you take Venezuelan equine encephalitis antiserum, say, or Rift Valley antiserum and you mix it with the virus, the virus's infectivity is neutralized. That is, if we take blood from a VEE survivor, separate the serum and mix it with some VEE virus, the antibodies in the serum will block the infectivity of the virus. That's the way we like to see things happen. This not only represents a weapon against the disease; the assay also serves as a functional measurement of the amount and potency of antibody present in the sample.

But if you take Ebola or Marburg virus and follow the convalescent antiserum procedure I just described, and then assay the infectivity, you're going to find that it's essentially the same. Both in vitro (in the lab) and in vivo (in the body), you don't get any significant neutralization of filoviruses with convalescent serum. Basically, this was telling us three things, none of which we wanted to hear. First, despite the isolation of the virus, we had no test to confirm the validity of results when we found a positive by the fluorescent antibody test. Second, it meant that antiserum likely provided no passive protection against the disease. And third, this was telling us there was something important about the virus that we didn't understand.

This continues to be a problem even today, and not just for

filoviruses. It is also true for other killers such as retroviruses, the viruses such as HIV that contain the reverse transcriptase enzyme which converts viral RNA into a DNA copy which is then insinuated into the host cell's reproductive mechanism. The big problem in battling HIV is that the virion is continually changing, which essentially means you're fighting today's HIV with yesterday's antibody weapon. Researchers are working with genetic engineering to try to get around this, but so far, we don't know whether or not it's going to work.

After the Zaire and Sudan Ebola epidemics, there were a couple of sporadic reports of new disease, the most significant of which was another outbreak in southern Sudan in 1979 with thirty-four known cases and a 65 percent mortality rate. CDC sent Joe McCormick out to investigate. One night, while taking numerous blood samples from very sick patients in a hut in Nzara, Sudan, he was stuck in the hand with a needle containing the blood of an old woman who was already symptomatic. Joe was understandably terrified that he'd inadvertently inoculated himself with one of the deadliest agents known to man, but boldly and admirably continued on with his work, figuring if he had been infected, there was nothing he could do about it anyway. Mercifully, the woman turned out not to have Ebola and Joe never got sick.

This is only one of many cautionary tales in the world of filoviruses and other hot agents. Fred Murphy left a major meeting on Ebola in Antwerp for a WHO consultation in Geneva. He became sick with fever and bloody diarrhea; when it was known that he had been working with Ebola with Guido van der Groen and had taken the electron microscope grids outside the plastic containment envelope to be able to stain them, Paul Bres shepherded him to a hospital where he remained in containment until he was found to be free of Ebola. The story of a Peace Corps volunteer working with the international team in 1976 is well known: his fever earned him a trip from Zaire to Johannesburg in a space capsule before it was discovered that he was not a filovirus patient. But lightning does strike, and Jeff Platt in England discovered this while working with the Sudan subtype of Ebola virus. He was injecting guinea pigs when he pricked his thumb with the needle. He immediately removed his glove and milked the site under a bleach solution to disinfect it. A close examination with a hand lens showed no penetration of the skin and he thought all was well until six days later, when he became febrile and complained of abdominal pain and a debilitating fatigue. He was

severely debilitated for nine days until he recovered. He received
several speculative treatments that were unlikely to have had any
effect on his disease process, and relatively few lab studies were done
because of the concern over having another case—but he had
squeaked by with a relatively mild case of Ebola.

Then Ebola dropped off the radar scope.

But several months after the Sudan outbreak, in January 1980, a
French expatriate who managed the pump house of a sugar factory in
western Kenya came down with Marburg after visiting the Mount
Elgon forest reserve in Kenya and Uganda. After developing the horrifying sequence of symptoms, he was flown to Nairobi Hospital, one
of the best in East Africa. And there he died, but not before infecting
a distinguished young physician named Shem Musoke, who lay near
death for weeks, but did eventually recover, giving this particular
episode a statistical kill rate of 50 percent. But that doesn't mean very
much when you happen to be the one who's got it. At that point
there are only two meaningful percentages: zero and one hundred.
Either you make it or you don't, and other than giving good supportive care, there isn't a whole lot the doctors can do for you to affect
that outcome.

In 1985, I made a couple of forays into the Central African Republic
as part of a team looking for hemorrhagic fevers. With me were Gene
Johnson, a civilian scientist at USAMRIID (no relation to Karl), and
two Frenchmen—Didier Meunier of the Pasteur Institute and Jean-
Paul Gonzalez of ORSTOM (Organization for Research Overseas).
ORSTOM was an interesting entity. It dated back to the days of
France's colonial empire, and, as far as I could tell, had been established essentially as a scientific means of determining how best to
exploit the resources of the overseas colonies. To that end, they did
soil typing, catalogued mineral deposits and animal populations, and
generally figured out what, where, and when they would pluck the
spoils of empire. In more recent times, the organization has taken on
a more altruistic aspect, and is now a first-rate scientific research
establishment. Gonzalez, who taught me an enormous amount about
central Africa, has become a major player in their virus research.

Driving in a Toyota Land Cruiser and two ton-and-a-half Toyota
diesel trucks and accompanied by six African helpers, we visited
small villages without any medical facilities at all. Wherever we went,
we looked for sick people, and when we found them, we would try to
find out what they were sick with—whether it was typhoid, malaria,

something transitory and less serious, or whether it was actually a hemorrhagic fever. We sought out physical clues, taking blood samples wherever we could, looking for antibodies even in healthy people. The antibodies could tell us if they had been infected previously with one of the viruses and recovered or perhaps had only a mild illness.

This can be a dicey proposition, fraught with practical problems and issues of medical ethics. The fact of the matter was that we were doing scientific research which we hoped would further our understanding of deadly hemorrhagic diseases such as Ebola and Marburg. But there was really very little of an immediate, practical nature we could offer these people in return for their cooperation that would be of direct benefit to them. You come into a village and ask to bleed people. For what? What good does it do them?

We tried to explain, through a chief or other tribal official, that we were there to find out what viruses and other diseases were around in their community with the hope that we might be able to stop them. But to do that, we had to have a sample of their blood. Sometimes we set up a little clinic and tried to treat whatever came in, but so many of the conditions were chronic medical problems. In some instances, we actually did blood typing on people just so they would have something tangible to show for their cooperation, though what a rural tribesman in Africa who had probably never seen a doctor before us is going to do with knowledge of his blood type, I really have no idea. The main thing was to show that we cared and were concerned with each of our "clients" as individuals.

How much cooperation we actually got depended on such variables as the prevailing local attitude toward outsiders and the relationship between a chief and the villagers. If the chief was influential and had the loyalty and respect of his people, then getting his agreement was generally enough. If there was tension and conflict between the chief and the villagers, then having that agreement could even be counterproductive. And of course, we couldn't always discern the village dynamics quickly enough to decide on the best strategy.

We began our expedition in Bangui, the capital of the Central African Republic, just across the Ubangi River from Zaire. We traveled through thick, lush forests, into the tree savanna, then to an area with lower trees, and finally north into the central African grasslands. We did this during the dry season. In the rainy season, aside from being wet and miserable all the time, we wouldn't have been able to get even our four-wheel-drive vehicles across the vast plains of mud.

For the first couple of days, we'd pass through larger villages that sometimes had a regional mission hospital or a clinic. We'd come upon an occasional market where we could buy bread and meat and supplies and perhaps even diesel fuel. They would usually have a generator which would supply power for a few hours a day. And we would sleep in French planters' compounds or religious outposts where we were often received as welcome guests to break their solitary lives.

But the farther we got from Bangui, the more remote and cut off we became. After about three days we passed through a game preserve and all the vestiges of Western civilization slipped away. There were no more markets, refrigeration, diesel fuel for sale, no more planters' compounds or religious missions. Just totally remote little villages.

I have to say the kids were one of the saddest things I've ever seen. They had virtually nothing—no books, and rarely any toys, not even a communal soccer ball. When you come from the United States or other parts of the Western world, no matter what your socioeconomic position, you have *things*. And you take them for granted.

We now slept outside near the trucks, on cots with French army sleeping bags and mosquito netting. It was like camping out in the woods as a kid, except for the lions roaring in the distance or mounds of elephant shit. This wasn't all that bad in itself since aging elephant droppings have very little odor, but you can't help thinking that with all this evidence of a lot of elephants, they could come back at any time.

There were some compensations, though. We lived under the French system, and that meant a full three- or four-course meal every night with wine, served on checkered tablecloths, prepared by our camp cook. Near the beginning of the trip, we'd sit down to a first course of soup made from freshly available ingredients, accompanied by locally baked bread and butter, chicken or some kind of fresh meat, expertly prepared with garlic or other spices, cooked vegetables, then perhaps a little salad, and finally a sweet or a French cheese and more bread or fruit. As difficult and demanding as the journey was, dinner gave us something to look forward to at the end of every day.

As I say, that was near the beginning of the trip. A week and a half into the expedition we were eating preserved bread, dehydrated soup, and canned meat, all of which tasted essentially the same. But we were still eating on checkered tablecloths and drinking wine (though the quality of the wine had dropped off considerably as well). The substance might have changed, but we always observed the ritual.

By the final week our supply of diesel fuel was getting too low to use on anything other than running the trucks and the once-fresh water in our fifty-five-gallon water tank was getting green and buggy. As to food, we were reduced to French army enlisted man's C rations: a biscuit harder than bone, accompanied by a can of cheese that had the consistency and taste (I'm projecting here) of wall putty. There was usually some kind of unidentifiable potted meat that was about 50 percent congealed lipid and utterly foul. We didn't have the right equipment with us to identify it microbiologically, so to this day I'm not sure what it was we were eating. And to finish off, the kit provided you with about twenty squares of toilet paper, which, if you didn't need to use it right away, could be saved to sand down your furniture back home. It reinforced my ironclad principle for traveling on the job: bring my own toilet paper. If you're going to put your ass on the line, treat it nicely.

My most memorable culinary experience of the trip occurred when the termites were swarming one day. The Africans had a technique for tapping the trees and bringing out a handful of them, or attracting them with lights, after which they would swallow them whole. Jean-Paul Gonzalez suggested I give it a try.

"Why not?" I responded. I collected a handful, quickly developing my own technique. I found that if you pop them into your mouth, chew well, and then swallow, the wings didn't tickle your throat going down. They had a delicate, nutty flavor and were so good that I started gathering them up and putting them in some of the plastic bags we'd brought for blood samples, which I'd then quick-freeze in liquid nitrogen, to be thawed out when we wanted a snack.

Like every other hemorrhagic fever fishing expedition before or since, we found no active filovirus cases, despite the samples we took, and no evidence of where the disease lived in its natural state. We did learn a few things, but it seemed as if everything we discovered deepened the filovirus mystery.

Among them was a stunner. Although we couldn't find the natural reservoir for this thing, if it just disappeared into the forest whenever it wasn't killing humans, why, now that we recognized the virus, were we seeing antibodies for it in the blood we were drawing and

testing throughout equatorial Africa? In some of the tribespeople we
tested in the Central African Republic, the positivity for Ebola antibodies ranged as high as 20 and 30 percent! So the first question we
had to ask ourselves was: were we incompetent? Were we missing
clues and evidence that were right there before our eyes? It isn't a
pleasant question, but it's one you always have to consider.

Was there a lot of disease out there that hadn't been reported? If
so, we should have been seeing great graveyards behind every village
in the endemic area. But we weren't. There was no evidence of ongoing viral hemorrhagic fever transmission. Conceivably, we could
have been out there at the wrong season or during the wrong year,
but there were no stories of pestilence, no reports from missionaries,
nothing to suggest this was going on. It didn't make sense.

So we wondered if we were seeing cross-reacting antibodies from
some other, as yet unknown filovirus, or maybe something else entirely. Were these people exposed to something that gave them antibodies that reacted to Ebola even if they had never been near Ebola
itself? Were these just meaningless artifactual reactions that we
couldn't check because of a lack of a neutralization or other confir-
matory test?

The idea of cross-reacting antibodies is not a new one, and, in fact,
cross-reacting immunity is central to the first effective vaccine, as
well as the eradication of the one killer virus ever completely eliminated from planet Earth. In the closing years of the eighteenth century, Edward Jenner, an English country doctor whose abiding passion was birdwatching, also observed that milkmaids who caught
cowpox, a fairly mild disease, from their bovine charges seemed to be
immune to the frequently lethal smallpox. Testing his theory that
exposure to the mild disease afforded protection from the devastating one, Jenner injected eight-year-old James Phipps with fluid (or
"vaccinia") from a cowpox pustule. Six weeks later he injected fluid
from a smallpox pustule. James did not come down with the dread
disease, and since then Jenner's innovation has become the basis for
vaccination and a monumental milestone in medical history, placing
Jenner forever in the pantheon, along with the likes of Louis Pasteur
and Alexander Fleming.

This interesting peculiarity of smallpox—that it is specific to the
human species but solidly blocked by a less serious animal disease—
caused world public health officials in the 1970s to believe that they
had a shot at completely eliminating it from the planet through a
massive, decade-long campaign of vaccination. And it worked. On

May 8, 1980, the World Health Organization declared the human species free of smallpox for the first time in history. The only remaining samples are frozen in liquid nitrogen at CDC and the Institute for Viral Preparations in Moscow (later moved to Novosibirsk in Siberia, without consultation). There is frequent talk of destroying these samples too. As a hedge in case this is done, the virus' genetic sequence has been mapped out for future reference.

It was hoped that the smallpox eradication campaign would provide a model for the elimination of other infectious scourges, and we are well along with polio and taking up the cudgel against measles. However, we are turning away from the notion that infectious disease can be relegated to a historical footnote. The more intelligent perspective, I feel, is that we will always have to coexist with the microbes, constantly resisting their incursions and their mutant flexibility. That is why I push the wisdom and utility of surveillance efforts so frequently.

If our team had discovered a smallpox-like coincidence with Ebola, it might have led to the first effective treatment. So unraveling the mystery we'd discovered was definitely a worthwhile undertaking. But we never did pin down anything useful, and, not seeing any more actual disease, we decided to roll up our program.

I went back to Africa just after the rainy season that year. I was determined to go into the hospitals and find those Marburg patients myself. Fred Feinsod, a young physician at USAMRIID, and I used Bruce Johnson's well-blazed trail through western Kenya, passing near the foothills of Mount Elgon. This was the "Bermuda Triangle" of Marburg. The monkeys that carried Marburg virus to Europe and notoriety in 1967 were exported from nearby Uganda, and the 1980 case was from the same area. Johnson had spent two years here looking for more Marburg virus without success. Again, we went around to local hospitals, this time in a VW bus, examining patients very carefully. We took samples from sick people and, like good detectives, conducted interviews and tried to retrace the steps of previous victims, determining what, if anything, they had in common. A big problem we found was that there was virtually no testing or surveillance for the common diseases like typhoid and malaria: the ones we refer to as the Big Rule-outs. And if we couldn't rule them out, then we really had no way of knowing how much filovirus was out there. At our behest, Peter Pettit, a Dutchman doing his master's in medical microbiology, returned the next year to do bacterial blood cultures on supposed viral hemorrhagic fever cases and found them positive for typhoid instead. I left Africa frustrated that once again I hadn't found anything positive about filoviruses.

There was one other filovirus occurrence in Africa in the 1980s. In August 1987, a fifteen-year-old Danish boy who attended a boarding school in Denmark went to this same area of Kenya for his summer vacation to visit his parents and older sister, who went to a private school in Nairobi. The parents worked for a Danish relief organization in Kenya. While on holiday in Kenya, the boy began to develop severe symptoms and was flown back to Nairobi Hospital for treatment. There, he came under the care of Dr. David Silverstein, an American physician working in Kenya who had treated the Frenchman and Dr. Musoke back in 1980.

As with the Frenchman, the Danish boy's condition grew worse and worse. He developed high fever, acute respiratory distress, and diffuse bleeding before succumbing to the disease in front of his anguished family.

The boy's blood and tissue samples were sent to USAMRIID by Bruce Johnson just as he ended three unsuccessful years chasing Marburg in Kenya. Gene Johnson isolated a virus and identified it as Marburg. Gene contacted Dr. Peter Tukei, a colleague who worked for the World Health Organization and was stationed at the Virus Research Center in Nairobi, and asked him to interview the family to try to find out where the young boy had been and where he might have contracted the disease. Gene had been working at VRC setting up a mini BSL 4 lab there, and knew Peter well.

A few days later, Peter called back with an intriguing bit of information. The boy had been exploring inside Kitum Cave, on the slope of Mount Elgon. What made that piece of information so frighteningly noteworthy was that seven years earlier the Marburg case who infected Dr. Musoke had also been inside Kitum Cave, a natural home for bats, mice, insects, almost everything small that slithers, walks, or flies, and a favorite haunt of elephants and other animals in search of salt and mineral deposits lodged in the cave's walls, as well as leopards that visit to snack off the others. The deadly filovirus might be living in something inside that cave. The question was, what?

It was a question that remains unanswered. Despite a detailed and meticulous scientific foray into Kitum Cave during which they collected literally thousands of samples, Gene Johnson and John Morrill found no positives for Marburg or Ebola. I was disappointed but not overly surprised. There were other possibilities for the common exposure of the two cases, even though the Kitum Cave was the best lead we had. It at least gave us a good look at an array of African fauna for signs of Marburg virus.

One thing the work did leave us all with was a healthy respect for what this mysterious microscopic virus could do. It caused acute suffering and one of the most unpleasant deaths imaginable. Marburg was bad enough; it killed about a quarter of its victims. But its cousin, Ebola, had killed nearly 90 percent. Ebola was the hottest of the hot agents, always investigated under Level 4 conditions. And even then, the space-suited scientists and technicians could not afford to become complacent or let down their guard. They were working with a killer that had no known cure.

9

Right Here
in River City

SO ALL OF THIS is why I was hearing the hoofbeats of zebras, rather than horses, that Monday morning, November 27, 1989, when Peter Jahrling and Tom Geisbert stuck that folder of electron micrographs in front of me.

Ebola. Right here in River City.

My mind was reeling.

If this really is Ebola and it's finally jumped out of hiding in a metropolitan area like Washington, D.C., we could be facing something none of us has ever seen before, something not experienced since smallpox days, or maybe since the great plagues of the Middle Ages. This one really could be the Andromeda Strain.

How did this thing end up in our laps?

USAMRIID did a lot of work with monkeys, since primates were the closest we could get to humans for medical testing. If we were going to send American soldiers into a hostile field where they might encounter chemical or biological warfare, we had to know whether what we'd given them was going to work. So thinking of Marburg and the often similar reactions of monkeys and man, back in April,

Peter Jahrling had asked me, if anything happens to come along which kills monkeys, should we be interested in looking into it? I said absolutely.

Things did come up from time to time and they could be very unpleasant. There had been an outbreak among monkeys at Holloman Air Force Base in southern New Mexico which spread from cage to cage. Five hundred rhesus and cynomolgus macaque monkeys had been lost. In an effort to save a few of the arriving infected monkeys, the caretakers had moved apparently healthy animals out of the quarantine room, which is exactly what you don't want to do since they could already be infected. As it turned out, they *were* infected, and the virus spread widely throughout the facility. Peter had worked that one up and isolated the virus. The killer turned out to be simian hemorrhagic fever (SHF), a horrible disease in monkeys, but one which does not cross species into humans. Still, as Peter astutely reasoned, anything that destroyed monkeys was a serious threat to our work and mission.

Simian hemorrhagic fever had first been identified during an outbreak at the National Institutes of Health in Bethesda, Maryland, in 1964, when it wiped out 223 monkeys from a colony of 1,050. The research work leading to the identification had been done by two distinguished virologists, Alexis Shelokov and Nick Tauraso.

Peter Jahrling had continued his involvement with SHF. In his spare time while traveling around the country to labs and research facilities in his official capacity, he would give presentations on the disease to try to raise the consciousness of primatologists and show them how to recognize the problem early on. That was why Dan Dalgard, a veterinarian for Hazelton Laboratories America, Inc., in Vienna, Virginia, and the consulting veterinarian for another company division, Hazelton Research Products' Reston Primate Quarantine Unit, called Peter when he realized he had a problem.

When monkeys are imported into the United States, they must undergo thirty days of quarantine to make sure they don't have any diseases before they're released to labs. This requirement was put into place as a direct result of the Marburg episode in 1967. Hazelton enjoyed a very good reputation in the research community. At the time, they were conducting contract studies on drug toxicity, carcinogenesis, and environmentally sensitive compounds. Dan Dalgard himself was universally respected. If he thought he saw trouble, there probably was trouble to be seen.

On October 4, a shipment of a hundred cynomolgus macaques

from the Philippines had arrived at the Reston unit and had been placed in Room F. By November 1, there had been a far higher number of deaths in that room than normal. Dalgard watched the situation carefully for several days. On Saturday, November 11, he conducted necropsies on the dead monkeys, and based on gross anatomy and the clinical symptomatology, he made a presumptive diagnosis of SHF. This really worried him. No veterinarian wants to see his animals suffering, and the loss of any one of them meant a substantial financial sacrifice to the company and/or a delay in a medical research protocol. He did some more necropsies on Sunday, then on Monday he called Peter Jahrling.

When Dan called Peter and sent over samples, Peter unfroze a supply of MA-104, a unique monkey cell line originally used at NIH for the first studies of SHF. He and lab technician Joan Rhoderick would introduce ground-up Hazelton monkey samples into these cells to try to culture SHF and get a definitive reading. This wasn't a guarantee. Sometimes you can get cultures in cells in vitro, but sometimes it has to be done in live monkeys. So Joan inoculated the MA-104 cells with the suspected virus samples, incubated them, then stained them with fluorescent antibodies from a monkey antiserum to SHF we had prepared in April.

By November 16, Peter reported that he had isolated SHF, and the Hazelton veterinary pathologist had found a microscopic picture suggestive of SHF.

Even before confirmation, Dan Dalgard made the difficult, but proper decision. He ordered all the remaining monkeys in Room F euthanized, health and security precautions tightened to the highest possible level, and no monkeys to be released from quarantine even after the thirty-day legal requirement expired.

For the next ten days, there were occasional deaths in the remainder of the Reston monkey population, but neither their pattern nor the pathology suggested SHF. It could have been "normal" attrition; monkeys do die in captivity just as they do in the wild. But Dan Dalgard was becoming increasingly troubled. His gut feeling was that something else was spreading in the monkey house.

Meanwhile, Tom Geisbert, the talented young electron microscopist Peter Jahrling had traded one of his personnel slots to obtain, was examining the Reston stuff Joan Rhoderick had cultured. He wanted to photograph it to see if he could confirm the diagnosis by finding SHF virions. But when he looked at the cultures under the regular microscope, it didn't look right to him. The fluid was milky and clouded, indicating floating dead cells. What was killing them? The remaining cells looked sick.

He took the flask to Peter, who glanced at it and concluded that some common bacterium like *Pseudomonas* must have gotten in, and that could certainly do the trick. He unscrewed the little cap on the end of the flask and wafted air toward his nose by waving his hand, looking for the telltale cabbage-like odor of *Pseudomonas*.

Surprisingly, there was no smell. Peter suggested taking a look under the far more powerful electron microscope to see if they could tell what was going on. So Tom fixed down a little bit of the stuff in the glutaraldehyde, then scooped it out and preserved it in hardened resin. It was the Friday before Thanksgiving and Tom was looking forward to a hunting trip in West Virginia. He'd be gone until after the holiday. He'd do the actual electron microscopy as soon as he got back to USAMRIID.

There were four cell cultures inoculated from monkeys that had died of disease in Room F that showed cytopathic effect—cell damage as a result of virus infection. The fluorescent antibody test had shown up positive for SHF in three of them. It was in the fourth sample that Tom found Ebola, and that was when he and Peter came to me. Once we knew what we were looking for, we were able to use other fluorescent antibodies to confirm the presence of Ebola antigens, possibly the Zaire subtype.

The worst of the worst. The pucker factor was extreme.

My greatest single worry at that point was how little we knew about what we were dealing with. The medical community has observed thousands of flu outbreaks in this century and we still don't really know how any individual one is going to behave. They all cause trouble, but 1918–19 killed 20 million people. By contrast, there was so little experience with Ebola Zaire, with all the filovirus strains put together, that no one could make an accurate prediction. This was ours because, as little as we knew, no one else knew any more. Most of what you did with Ebola was go to Africa and count corpses after the fact.

Adding to the uncertainty was the fact that RNA virus molecules do not always pass exactly intact from one generation to the next. The army first found this out with its dengue fever vaccine. RNA-based viruses tend to be not terribly careful or accurate replicators, compared with their DNA cousins. Each time an influenza virus enters a host, for example, what comes back out can be slightly different from the strain that went in. You do this enough times (and compared with animals, viruses replicate exceedingly fast), and you can end up with something substantially different in terms of biological properties and antigenicity from what you started with. This is one reason there are so many strains of flu. It's also why having the flu doesn't immunize us for the next strain that comes along—the influenza virus antibodies we've each developed don't recognize the new strain because it's not antigenically similar enough to cross-react.

With the flu you're miserable for a week and then you get better. But we were dealing with Ebola. With Ebola you're miserable, then you're worse, and then, oftentimes, you die. What if each time Ebola comes out of hiding it passes along a slightly different genotype into the human population? If so, would this particular genotype be more or less deadly than Ebola Zaire, than Ebola Sudan, than Marburg?

Even if we had sequenced the entire virus genome at that point, and we hadn't, all that could have told us was whether there were any differences between this and previous strains, not what those differences meant. The only way we could tell that is through what we call the "feet up or feet down" method. You go check on your infected hosts and if they're alive they're standing there with their feet down on the floor. All we had to go on here was a lot of monkey feet up, and our people correlation several years ago was a lot of human feet up in Zaire.

Keep in mind that when something like this turns up on your plate, the rest of your world doesn't suddenly stop. We all had full-time jobs and were trying to deal with all the routine, day-to-day stuff. Suddenly, though, you find your priorities have changed. The military is mission-oriented, and we now had a mission.

But the results were ambiguous. Among the benefits of genetic research has been the production of a particular molecular derivative of the immune system's B cells or B lymphocytes, known as a monoclonal antibody. This is actually a complex concept in biology, but here's the thumbnail: the B lymphocyte is responsible for recognizing an antigen once it has entered the body. We are all born with a large number of individual, distinct B-cell lines and make even more of them to be ready for future calls to action; we call this our *B-cell repertoire*. The antigen stimulates individual members of the B-cell repertoire to make more identical B cells and to mature into a form that secretes antibodies to battle that antigen. A particular antigen (we hope) will activate many of those B-cell lines in defense against the intruder. So if, say, we have antiserum to a particular virus—let's say for Rift Valley fever—what we're really saying is that we have *polyclonal antiserum*; that is, the Rift Valley antigen will stimulate only a certain subset of the B cells (the ones which, for genetic reasons, can recognize this particular antigen), and they, in turn, will proliferate and produce the proper antibodies. So you might have dozens or hundreds of separate B-cell lines that respond to Rift Valley, and they are not all the same.

What a monoclonal antibody is, then, is the product of a particular B cell that we can immortalize and clone through modern genetic engineering, which means it is very specific. It's not just regular old antiserum that contains polyclonal antibodies to Rift Valley fever; it also carries antibodies to God knows what else—virtually anything the donor has had and recovered from.

To a molecular biologist, this is quite neat. The only problem is that these monoclones don't come with tidy labels that tell you exactly what they're good for. You have to characterize each individual one you've produced to see exactly how it reacts. It is a tedious and arduous process and you have to have a lot of known quantities to try to tag each one. We probably had hundreds of different monoclones for Rift Valley fever, probably thousands for various flu strains. We had about a dozen for Ebola, all derived from 1976 isolates of Ebola Zaire.

So we threw the small barrage of Ebola monoclones at our Reston samples. If they reacted, then we were dealing with the same disease; if they didn't, it was probably still a filovirus, but we wouldn't be sure exactly what. We held our collective breath as we waited for the result.

As it turned out, most of the monoclones reacted, but some didn't.

I wasn't sure what this meant, but I knew it could be a problem. We could get a reaction of either "Well, don't worry, then, this isn't really Ebola!" or "This must be a new strain of Ebola and the sky is falling!" We'd probably get both reactions at the same time. We had to learn more, and we had to learn it fast.

In managing a potential medical crisis, you have to know when to bring other people on board. Timing is critical. If you start sounding the alarm bells too early, you're going to look stupid and blow all your credibility if it turns out to be nothing. If you wait too long, you will be behind the power curve, and it'll also look like a cover-up.

In January 1976, just before Ebola descended on Karl Johnson and Special Pathogens like a ton of bricks, an eighteen-year-old army private named David Lewis collapsed and died just hours after completing an all-night hike at Fort Dix, New Jersey. Before the march, Private Lewis had complained of feeling feverish, nauseated, dizzy, weak, and achy: flu-like symptoms. He'd been advised not to go on the hike, but he was a soldier's soldier and didn't want to miss it.

Early each year, the flu watchers begin looking for the influenza strain they think will provide the most serious threat, and prepare a vaccine for it. Normally, this vaccine prevents about 70 to 90 percent of flu cases and is particularly recommended for the elderly, anyone with a compromised immune system, some health-care workers— anyone for whom getting the flu would be a bigger problem than simply a week or ten days in bed. But every once in a while, a really bad strain comes along. Really bad, of course, is a relative term. But in 1918, there was a worldwide flu epidemic that was really, really bad by anyone's definition of the term. It killed a half million people in the United States and twenty million throughout the rest of the world. If you look at the year-by-year mortality tables for the twentieth century, nothing else comes close to the 1918 flu—not the trench warfare of World War I, not the carpet bombing or wholesale civilian slaughter of World War II, not Stalin's purges, not any other natural disaster, nothing. And around the American Bicentennial year, many of the experts thought they were seeing signs of another 1918-type strain, which was called Spanish flu because the Spanish were one of the first to admit they had it, or swine flu because its infectivity had been proven through cultures grown experimentally in the nasal passages of pigs and then introduced into other animals. The evidence suggested that the previously hale and hearty Private Lewis had died of swine flu. Where he got it, no one knew, but if this was the index case, there was still time to act. Very few people still around would have antibodies from swine flu, and if you extrapolated the 1918 death toll to 1976 population levels and factored in air travel, the world was looking at a potential nightmare.

That was why, with the backing of a whole slew of infectious disease experts, from CDC director David Sencer to the twin immortals of polio eradication, Jonas Salk and Albert Sabin (in an extremely rare show of public agreement), President Gerald Ford boldly ordered a rush vaccine-production effort and a massive swine flu vaccination program.

Which was done. The only hitch was, there was no swine flu. Private Lewis's death from what was officially termed A/New Jersey/ 8/76 influenza turned out to be an anomaly. It was probably the combination of his flu, bad luck, and the rigors of an all-night march in freezing rain that did him in. All that health authorities had to show for their efforts was a lot of money spent, a lot of panic stirred up, and a handful of people who had contracted paralyzing Guillain-Barré syndrome as a side effect of the vaccine, leading to much needless suffering and another $93 million paid out by the government in claims.

And all from one questionable case. Even in retrospect, it was difficult to say it was the wrong decision. USAMRIID's ultimate boss, General Philip Russell, M.D., head of the U.S. Army Medical Research and Development Command, a very good virologist and one of the brightest and ablest people I've ever had the pleasure of working with, had seen no choice but to go ahead and recommend to the Army Surgeon General the vaccination of military personnel once he'd reviewed all of the available data.

That's what was flickering through the back of my mind as I closed my office door and stared at Tom Geisbert's electron micrographs of cultures from a single monkey who apparently had Ebola. That was all of the available data this time. I tried to shake off the epinephrine surge, collect my thoughts, and filter out all of the emotional static.

Peter redid Joan's fluorescent antibodies, just to be sure, while I went home, where I pulled out a couple of Ebola reprints to refresh my memory. I gave my wife, Susan, some sense of what was going on, but she didn't have the scientific background to grasp the import of what we had found. She could tell, though, that I was deeply concerned. She recalls my saying, "We're working with something which, if it turns out to be what we think it is, is gonna make a lot of people crap in their pants. Or maybe it's nothing. We just don't know." What I do know is that for the duration of the crisis I was getting up at five or five-thirty every morning, racing out the door with my motor already running full blast. Anyone who knows me well knows it's all I can do on a normal workday to pry myself out of bed by eight o'clock. Left to my own devices, I don't really kick into high gear until close to noon.

Who had to know? I started asking myself. Well, first of all, Colonel David Huxsoll, USAMRIID's commander. Then General Philip Russell; it would be up to Phil to decide how far up the chain of command this thing went. Surely it would have to go to the Army Surgeon General, and from there most likely to the Joint Chiefs and Secretary of Defense Dick Cheney.

Then there was the question of CDC. They were the ones who had primary responsibility for public health. At USAMRIID, it never occurred to me that I'd be involved in something like this here in the United States. At the moment, this was a monkey problem and we were going to do our best to keep it that way. We had the samples, we worked with monkeys in the hot suite all the time, and we were confident we could do a better job with them than anyone else. CDC had the national responsibility for human health and had to be told promptly. The state of Virginia had the immediate responsibility for everything on their soil. I decided we would call in CDC and the state as soon as we had sorted things out with Phil.

I wish I could say that the issue of turf didn't enter into the equation, but that would be a lie. As I've noted, turf wars of one sort or another enter into practically every equation in our business. We're all jockeying over the same real estate, sometimes to occupy it and sometimes to get rid of the responsibility for it. And in the galactic scheme of things, I can't say any of us has any more right to it than the viruses and bacteria, the cockroaches, or the elephants that left us their calling cards in the savannas of central Africa. And much as we taxpayers would want our various government agencies to coexist peacefully to the benefit of all, the reality is that the army has conflict with the navy, the FBI has conflict with the Secret Service, and the CIA has conflict with the Defense Intelligence Agency. I was determined USAMRIID would neither be pushed aside nor tread on the legitimate turf of the Centers for Disease Control. What we hope is that in the case of a real threat, the key players will be working on the same team.

But Joe McCormick, the chief of Special Pathogens, was my opposite number at CDC, and our conflicts over the years were as well known in the public health community as our basic respect for each other's talents and track record. We'd spent years hunting this thing, and I'd be goddamned if I was going to turn the entire operation over to Joe or some folks in Atlanta, many of whom I'd never met, who'd have to start from scratch.

We worked through the next day, November 28, at the same time trying to get more answers from the lab and formulating plans for each possible contingency. I called Dave Huxsoll and told him we had something really big and we would need an open slot with him later on. Then I called General Russell's office. He was out, but I told his secretary that we would be over with Colonel Huxsoll and that we needed time on his calendar.

One of the key questions was: Who might already have been affected? There were probably people at the Reston facility who'd

handled the monkeys, like the people in Marburg, and they'd likely
had a chance to become sick and infect others. That was a job for
CDC.

We had gone over the accumulating data again. All the initial tests
had been repeated. Tom Geisbert had gone to the actual monkey
tissues and confirmed the presence of Ebola virions within the tissues
themselves. Our pathologist, Nancy Jaax, a veterinarian with a lot of
lab experience with hemorrhagic fevers, had confirmed Peter's diagnosis from autopsy slides. Nancy had done some work with Gene
Johnson, and she and John White, our veteran pathologist recruited
from behind the fence, had actually looked at Ebola in monkey tissue. Nancy wasn't the only Jaax at USAMRIID. Her husband, Jerry,
who was also a lieutenant colonel and also a "doggy doc," was chief
of RIID's veterinary division.

Peter called Dan Dalgard and told him that there was definitely
simian hemorrhagic fever in the monkey colony. But he added,
"There's also something else in some of those cultures."

"Well, what is it?" Dalgard asked.

"We're working on it," Peter said, "and in the meantime you and
your people should be real careful. There are serious potential public
health hazards if it turns out to be this particular agent."

Even though Peter was being cautious until he had more facts, Dan
picked up on his concerns and asked him whether it was Marburg or
something similar. Reluctantly, Peter conceded that was what he had
in mind.

He, Nancy Jaax, and I met with David Huxsoll in his office in the
main USAMRIID building. Dave is a D.V.M., a Ph.D., and a noted
rickettsiologist. We told him what was going on and he went through
the standard questions, just to assure himself we knew what we were
talking about. But I think he knew we did.

"I'll make an appointment with Phil Russell," he said.

"We already have," I said.

The three of us plus Huxsoll trooped over to General Russell's
office. He was just back from a meeting at the Pentagon. He and I
were good friends and in one-on-one or social situations I called him
Phil. In a public meeting I would always call him General. Aside from
being a first-rate physician and scientist, he had the political savvy
you need to work your way through the system and be an effective
general, yet he would never sacrifice anyone under his command for
his own means.

"General Russell," I began, "I think we've isolated Ebola virus from a monkey quarantine facility in Reston, Virginia." There's no sense beating around the bush with the guy who's got to decide what to do.

Peter gave Russell all the data, which he'd organized into tables. Nancy confirmed everything Peter said.

Then Russell got right to the heart of the action plan. "Who have you told?" he asked.

"Nobody," I said. "We wanted to have a coherent strategy first." I might have added that under no circumstances in the military do you want to blindside your boss, but he knew that already.

"We'll have to notify CDC," I stated.

For the first time in the meeting, I saw Russell smile. "Yeah, that's gonna be fun. Who else?"

We made up what we considered to be a responsible list, and the list was daunting: CDC had the health responsibility for human be-ings—Joe and of course Fred Murphy, as director of the Center for Infectious Diseases, would be at the heart of the storm—and quaran-tine activities came under the CDC's Center for Prevention Services. The U.S. Department of Agriculture supervised caged animals. The Fish and Wildlife Service was responsible for nonhuman primates. The Food and Drug Administration regulated monkeys used for vac-cine research and production. NIH oversaw the national monkey supply for the scientific research community. The Virginia State Health Department would have primary authority over the whole thing, and, technically speaking, CDC couldn't set foot in the state until state health authorities asked them to. And, of course, Hazelton owned the monkeys, so they had as strong a stake as anyone.

After we'd talked strategy for a while, Russell said to me, "What do you want out of this?"

"Let CDC take the public health issues," I conceded. "But the monkeys—we have to do that. We're the only ones who can do it safely, plus there's important scientific information to be learned— stuff we've been trying to find out for years."

The meeting ended with General Russell officially placing me in charge of the operation. This reminded me of the old story about the guy who's tarred and feathered and ridden out of town on a rail; when asked about the experience, he replies, "But for the honor, I'd just as soon have walked."

On the evening of November 28, Phil Russell called Fred Murphy, head of CDC's Center for Infectious Diseases, at home in Atlanta. Fred said he'd be on a plane first thing the next morning.

Nancy, Peter, and I divided the task of notifying the rest of the usual suspects. It was amazing how many people had the exact same word-for-word response and inflection: "Holy shit!"

Peter and I arranged a conference call with Dan Dalgard at home and told him there was serologic and electron-microscopic confirmation that the second agent found in the monkeys was Ebola virus, and that I felt obligated to notify national and state health authorities. Dan said he would alert Hazelton officials.

The authority I most wanted to talk to was the Virginia state epidemiologist, but I couldn't seem to get in touch with him or her. Every number I called either did not answer or gave me a recording. Finally, about seven in the evening, I called the University of Virginia Hospital in Charlottesville, asked to be put through to the emergency room, then asked for the resident on duty.

When he came to the phone, I said, "Look, I need your help. This is Colonel C. J. Peters of the United States Army Medical Research Institute of Infectious Diseases at Fort Detrick, Maryland. I have a potential health crisis. I need to get in touch with your state authorities, but I can't raise anyone."

Sobered, the resident looked on the wall list and gave me a couple of other office and home numbers, but still no one answered. I made another round of calls, stressing the urgency of the situation on the messages I left, and gave my home phone number in Frederick.

I was in my office later that evening when Peter knocked and stepped in.

"C.J., I need to talk to you about something."

"Okay."

"Before we had any idea this would be a filovirus, Tom and I were looking at the flask. I took the cap off and passed it near my nose to show Tom you could smell bacterial growth, that some bacteria had characteristic odors. Tom smelled it too."

I was stunned. "Peter, why did you do that?"

"Well," he said, "I'm not sure why. I thought it was bacterial. And now I thought you'd want to know about it."

In practice, bacteriologists whiff and sniff their culture plates all the time, because many bacteria do have characteristic odors. But bacteriologists are dealing with big, sticky particles growing on a petri dish. This isn't something anyone working with aerosol-infectious viruses that are potentially serious should ever do. Peter was one of the most careful and conscientious scientists ever to step into a

Level 4 lab, and he had tremendous experience with hot agents. I was really surprised by this lapse.

I had a decision to make. Did I throw him and Tom Geisbert into the slammer and wait to see what, if anything, happened? I knew Peter certainly didn't want that, particularly with the action heating up. But I also knew he trusted me enough to accept my ultimate decision. He knew I wouldn't make my decision based on friendship. If I thought these guys might be infected, then he knew that's where they belonged, in quarantine

"Did you get it on your hands?" I asked.

"No, we had gloves on."

"Did you shake the flask?"

"We picked it up but we didn't overtly disturb it."

"Did you waft it over to your nose or did you sniff it, or did you shove it up one nostril?"

"We kind of wafted it over."

I thought about it some more. We have this commitment to the community, and that's one of the reasons we set up the slammer. But we'd come to understand through our research that there doesn't seem to be any virus excretion before you're sick. Fever corresponds to the onset of real disease. It's not something I could prove beyond a shadow of a doubt, but we were seeing more and more evidence. If I'm right, even if those guys aren't infected and they did inhale some virus particles, they're not going to be infectious for others until after they're febrile. So we could watch them and take action if necessary.

"With what we know about barrier support treatment in Africa, I think your chances of having gotten it are very small. The community is going to suffer if we don't have you and Tom working on this," I said. "So I'm gonna have you both take your temperatures twice a day, and if anything changes, you let me know immediately."

Peter was obviously relieved. But he had a loving wife and three wonderful kids at home. If Peter or Tom was infected, I would put them into the slammer at the first sign of any symptoms. I knew, though, that there would be a lot of crap coming from every direction over the decision.

Early the next morning, which was Wednesday, November 29, we heard back from everybody we'd called the day before. We set up a meeting for that afternoon at two o'clock at Detrick's headquarters building. At the meeting were Dan Dalgard and two or three other people from Hazelton, Virginia health authorities, and representatives from all of the affected agencies. There were military and Hazel-

ton lawyers there too. The military folks were all decked out in green
and included representatives from Medical Research and Development Command (our parent organization), the Fort Detrick commander (our landlord), and of course USAMRIID. Altogether, there
were somewhere between forty and fifty people in the room, all
assembled on very short notice.

This was the first time I'd met Dan Dalgard face to face. In his mid
to late fifties, he was a distinguished-looking man who, like me, also
happened to be from Texas. I don't remember if he actually wore
cowboy boots, but he should have, and he certainly does in my
imagination.

I started the meeting off with a scientific presentation to bring all
the parties quickly up to speed. Peter gave most of the data, assisted
by Nancy Jaax. Peter passed around the electron micrographs so that
everyone could see what we were talking about.

When they reached Fred Murphy, probably the world's preemi-
nent viral electron microscopist, he looked closely at each one. Everyone held their breath awaiting Fred's pronouncement.

"This is real, Peter," he declared. "Congratulations. This is a good
piece of work." The sage had spoken.

One of the things I stressed was that we had to find out exactly
where these wild monkeys had come from and try to figure out how
they'd been infected. We had to try to find the source of the virus.
Otherwise, we were just waiting for another time bomb to explode.

During a break in the meeting, Phil Russell, Fred Murphy, Joe
McCormick, and I got together to hash out who was going to do
what. I brought up my suggested division of labor with CDC. "You
guys do the human epi, we do the monkey epi."

Joe didn't seem too happy to share the action, but Fred readily
agreed.

There was no question of the tension between Joe and Peter, Gene
Johnson, and me. To me, Joe's attitude was: I'm from CDC, thanks,
guys, but this is now my show; Gene and C.J. have hunted Ebola in
Africa but I'm the only one who's actually seen it there in human
patients, so I'm taking over from here.

No one who knows anything questions Joe's skills and experience.
And no one in their right mind could doubt his courage. During the
1979 Ebola outbreak in southern Sudan, he volunteered to go alone
into that war-torn country to collect human samples. Even the local
pilots who ferried him into the hot area were afraid to stay with him.
Using only simple paper and latex barrier protection, he collected his

samples by flashlight from dying people under the most primitive conditions imaginable. And after he'd jabbed himself with the needle, he continued working through until his mission was complete. We all respected Joe for those attributes. But what we couldn't get past was our perception of his arrogance and self-righteous attitude that he knew better than anyone else. He had concluded that this thing was not as catchable as some of us feared. Well, maybe it was and maybe it wasn't. The results of Gene's aerosol experiments didn't give us much reassurance on that score, and as far as I was concerned, I wasn't going to take any chances with my people until I knew more. The Russians have since published additional aerosol experiments that heighten my concern for this aspect of filovirus spread. I think we also bristled at his denial that we had a part to play in this. It was clear to me that it was going to go better for everyone if we worked together rather than at cross-purposes.

Dan Dalgard asked, "How long does it take you to test a monkey?"

"Using virus isolation, about a week to ten days," Peter replied.

"There's not much experience with these viruses," I added.

Joe McCormick offered, "We can get the answer back in two hours with our nucleic acid probe test."

I wondered out loud how they were going to pull this off since no one knew the genetic sequence of this virus to design a sensitive probe, but at that point, the group wasn't much interested in details. We had the big picture to worry about.

Fred and Joe had to get back to Atlanta. Joe wanted to take some of the virus back with him. Peter bristled.

"Joe, you've got to understand, we just isolated it—we don't know if there've been any contaminants."

But Joe wanted it, and as long as he understood the fine print, I thought he should have it. So I asked Peter to get a sample out from the Level 4 lab and pack it up, which he did.

The problem at Reston wasn't resolving itself. In addition to the Room F involvement, twenty-nine monkeys in Room H had also died. The disease, despite Dan's effort to isolate Room F, was spreading. I didn't know whether it was Ebola killing the monkeys, SHF, or something else altogether, but whatever it was, it was making them bleed and die. This was particularly distressing because Room H represented a separate shipment of monkeys that had arrived on November 8, more than a month after the Room F animals.

At this point we knew there had been both SHF and Ebola in the

Room F monkeys, suggesting it was brought in by one or more of them, and there was something deadly killing the monkeys in Room H. If that turned out to be Ebola too, then there were two possible explanations. Either some more of the virus came in with the November 8 shipment, or the Room H monkeys picked it up from the infected Room F monkey or monkeys. Like most commercial buildings, the Reston facility had common ducting from room to room. We'd done studies in the Level 4 lab which suggested that under the right circumstances, Ebola was transmissible by aerosol, and certainly by droplet. If this was being transmitted through the air, we were in real trouble.

Nancy and I headed out to the Hazelton-Corning headquarters on Route 7 in Virginia, near the suburban sprawl of Tyson's Corners. Although we were in our uniforms, we drove our own cars rather than military vehicles in order to avoid attracting attention.

The Hazelton offices were in a complex of modern two-story buildings that created a mini-campus atmosphere. Dan Dalgard met us and showed us into the lab, where one of his veterinary pathologists had prepared tissue slides. Nancy sat down at the microscope and studied them. At this magnification you couldn't see actual virions, but the cytopathic effect she saw was consistent with what she'd seen Ebola do in the lab. It was one more clue leading us toward the killer scenario.

Meanwhile, I asked Dan about needle use on the monkeys. We knew that Ebola had been spread in Yambuku by needle reuse without sterilization. Dan told us that although it was his policy not to reuse needles, he couldn't swear that it was always followed. Syringes were recycled after injections. Also, there had been times when inoculations had been given from multi-dose vials, so perhaps there could have been some transmission that way.

We arranged with Dan to take back dead monkeys from Room H and a few from other rooms. He called a guy at the Reston facility and told him to prepare the corpses for us, then went over there himself. Nancy and I had ridden over together and were told to meet this guy at an Amoco station on Route 7 near the entrance to the town of Reston, which would be on our way home.

We found the gas station and waited in my car—an early 1980s Toyota Corolla, a kind of burnished sienna red in color. Nancy and I talked some about what we were going to do with the monkeys once we got them back to the lab. I could tell she was more than a little nervous.

It was about four-thirty in the afternoon, the sun was going down, and the notorious northern Virginia rush hour was in full swing. There was a 7-Eleven next door to the gas station and I noticed that well-dressed people kept pulling up, getting out of nice late model cars, and going up to use the pay phone in front of the store, which struck me as odd. These people were all close to home, and some of them probably had car phones—what was going on? Unless they were all doing drug deals, which I thought unlikely given the surroundings and the clientele; they must be cheating on their spouses, I mused, arranging secret assignations out in the suburban heartlands. Life goes on, I reflected. I wondered if these people realized how quickly their lives could change. I thought back to Cochabamba, to Donato's deathwatch and the poignant meeting with his grieving young wife.

The guy from Reston finally showed up in a light green company van. He opened up the back. There, lying on the carpeted floor, were five frozen monkey carcasses, wrapped in paper towels and then in three layers of green plastic garbage bags.

"Well, here they are," the guy from Reston said.

Great. These monkeys may be frozen, but if they've got Ebola, they're still hot as hell. And we're supposed to schlepp these things back to Detrick like this?

"What are we gonna do with this?" I asked out loud. Clearly, someone as experienced and responsible as Nancy, used to taking every precaution known to science in the Level 4 suite, was not real comfortable with the haphazard way this operation was evolving. But in addition to being a good scientist, Nancy is also a good soldier.

"Sir, it's your decision if you want to transport them," she said.

I thought about it. Not accepting them in these circumstances would be the best way of covering our institutional ass. But that wouldn't get them where they needed to be or get the job done. Shit, I thought, anything that could happen to me could happen to him if I send them back.

The five frozen monkeys fit fairly neatly in my trunk, which fortunately was reasonably clean, since I'd thrown most of my junk onto the back seat. Just to be sure, though, I made a quick search for sharp edges. I didn't want any of those bags puncturing. Nancy, who had gotten a tear in the glove of her space suit one time in the Level 4 lab back in 1983 when she first started handling Ebola and nearly breached her protection, was highly attentive.

And I drove back to Fort Detrick very, very carefully. I figured if I was stopped, at least no cop would have wanted to confiscate my cargo as evidence.

When we got back to USAMRIID, Ron Trotter on Nancy's staff came to take care of the animals and we went up to the conference room of the Disease Assessment Division, which had become our command post. We now had some month-old tissue blocks, plus these new fresh-frozen monkeys to work with.

There were too damn many variables. That's what drives the panic factor in the general population, and concerns the scientists making critical decisions based on incomplete data. The previous experiences with Ebola had been in rural African settings. The people had lived in huts, needles were used repeatedly, hygiene was minimal, relatives made a habit of washing and handling their dead. But here in the United States, there was more mobility among the population, more opportunity for spreading the disease. Contacts between those infected were more anonymous and wide-ranging. Our buildings were climate-controlled. The mid-Atlantic region had a temperate climate, not a tropical one. We didn't know what the final sum of implications of this were for transmission.

Nancy and Ron Trotter went into the Level 4 lab to dissect the monkeys and examine their internal landscapes. I say "went into," but it's a lot more complicated than that. First, before you can even be authorized, you have to be intensively trained and you have to receive inoculations against anything potentially inside that we have inoculations for. Next, you go into a locker room and take off all your clothing, even your underwear. Then you put on a clean scrub suit and surgical gloves, followed by the full-body "space suit," manufactured by Chemturion and called a blue suit to distinguish it from the lighter-weight orange Racal suits used in biohazard field work. Then you both check for rips, tears, or anything that doesn't seem quite right. Only then are you ready to go through the last air lock into Level 4: the best containment lab human beings have been able to come up with for diseases for which we know of no cure.

When we see the space suits with their large angular hoods and clear plastic faceplates, we tend to think of an undersea sort of world of eerie quiet and serenity. But this is not the case. Once inside Level 4, you hook up your air hose to one of the overhead supplies and you immediately hear the din of rushing air all around you. Communication with your lab partner and speaking on the telephone are difficult, sometimes impossible. If you must speak, you may have to momentarily turn off your air supply, which is not a comfortable feeling. The environment within the hot suite is so claustrophobic and the noise within the helmet so loud and intrusive that some people can never get used to it.

Nancy and Ron slit open the monkeys and took a look around. There was some sign of hemorrhaging, but nothing gross or overt enough to confirm either SHF or Ebola. Making sure not to cut themselves (or each other) or puncture the protective barrier separating them from the virus, they began snipping off samples of the internal organs and placing them in vials for further study—empty vials for virology, formalin-containing plastic jars for pathology, glutaraldehyde for electron microscopy, a jelly-like compound for frozen sections.

Our focus now was on the monkeys in Room H. There had been a few deaths in the first few days, which Dalgard attributed to dehydration, the stress of travel, and the usual things that generally take out a few animals in each shipment. Then the mortality quickly leveled off. But around days fifteen and sixteen, the monkeys started dying again. Twenty-nine animals had died from this shipment so far. We couldn't interpret all the deaths because we didn't have all the corpses to examine. But from the samples we did have, the studies pointed to all the later deaths as coming from Ebola. As we'd feared, the sacking of the monkeys in Room F had not meant the end of the outbreak.

Dan Dalgard once again made a tough decision. He asked us to come in and destroy the remaining sixty-nine monkeys in Room H. We were the ones best set up to do it.

I agreed with the decision, but it would be a complex and dangerous operation. The Reston monkey house was a hot zone, just as the hospital in Yambuku had been, as Kitum Cave may or may not have been. I decided that if my people were going to be working in there with monkey organs and monkey blood, then they were going to have the same protection they had in the hot suite. I was not sending my troops into harm's way without the greatest possible protection.

Jerry Jaax was in charge of animal resources at USAMRIID; I asked him if he would take charge of the operation. Spiritually, I think Jerry is at least as much a soldier as a scientist, and he agreed as soon as I told him the commander had signed off on the operation. He then went to talk to Gene Johnson, who had handled the biohazard exploration of Kitum Cave and had more field experience with this sort of thing than any of us. The two men began plotting their strategy. The key priorities would be safety, containment, dispatching the monkeys as humanely and painlessly as possible, and bringing back samples. Never far from our thinking was the Andromeda Strain scenario, with images of the White House, the Capitol, and the Washington Monument superimposed.

On Thursday, November 30, Jerry and Gene drove over to Reston to do a little recon and finish planning the operation. Later that morning, Nancy Jaax and I drove over to Hazelton's office to meet with Dalgard and the Virginia State epidemiologist. Joe McCormick, back up from Atlanta, arrived about the same time with Steve Ostroff, a young CDC epidemiologist. They were going to go through the staff medical records and set up a surveillance program on all the people who had had any contact with the monkeys. They were also going to track which outside labs had received monkey samples for routine testing, so they could trace back who had had close contact.

One thing Joe had decided and I agreed on from the beginning was that there should be no attempt at a quarantine. Not that the military would have any role in either the decision or the execution. It simply doesn't happen as in the movie *Outbreak,* which some mistakenly thought was based on the Reston drama. For one thing, an army quarantine would violate the doctrine of *posse comitatus,* forbidding us to undertake military operations against civilians within our own country. But even on a more practical level, our experience has shown that quarantines don't do much good, because the resourceful can easily get around them, not to mention the press and public furor. It was true in Zaire, and it would certainly be true here. And we were both convinced that the best policy was simply to identify people who might have been exposed, explain the situation to them, and enlist their cooperation. If anyone "got hot," they would have to be quickly evaluated because they might be on their way to becoming contagious.

One thing Dan was concerned about was the potential medical and psychological effect on his employees, specifically the ones who had handled monkey samples. He collected a group of the workers, and Joe and Nancy spent a fair amount of time talking to them, trying to reassure them. When I spoke to Nancy afterward, she felt as if there was only partial success.

Dalgard was as interested as anyone in finding out what was going on with his monkeys. Joe didn't have the results from the two-hour probe test, so Dalgard turned from Joe and back to me. "Okay, Dr. Peters, let's figure out what we're going to do with the monkeys."

In fact, we were well on the way to having a good test for immediate diagnosis of Ebola-infected monkeys. After our meeting with all concerned, we had returned to the Disease Assessment Division office area and I pounded a couple of filing cabinets in frustration. Tom Ksiazek obtained some of Gene Johnson's reagents made for his epidemiological studies and went into the BSL 4 lab with the monkey samples that night. He focused all his concentration on the problem and went through a series of tests patterned after those he was using successfully to diagnose other viruses—and which would become the tests we use as the first line of defense against hemorrhagic fever viruses today. By dawn he had shown that he could put together a prototype ELISA test that would detect Ebola viral antigens in tissues or blood from seriously ill monkeys. Later, he would refine and extend the test to monkeys with milder disease, and finally in 1995 it would be the test we used to make the first diagnosis of Ebola in humans in Zaire. Tony Sanchez at CDC later obtained genetic sequences from the virus strain in Reston and developed a PCR test that, while more time-consuming than the four-hour antigen-detection ELISA, was even more sensitive.

Dan took us down Route 7 into Reston for our first visit to the monkey house. It was a square, one-story, faced-cinder-block structure in a typical smallish suburban industrial park—large parking lot in front, trees in back, and down a gentle hill from the trees a daycare center where the working moms of Reston deposited their kids at the beginning of the day. There was a McDonald's down the street that ran beside the parking lot and next to it a Taco Bell. If I were writing a fictional disaster movie scenario, that's just what I would have done: put a bunch of adorable kids and a fast-food restaurant right at ground zero.

As we entered, I was immediately aware that no one was using any respiratory protection. The risk was theoretical and in the office area would even be small. I glanced at Nancy and we decided not to make a big deal of it. I just wanted to move this thing forward.

As we entered the building, we were hit immediately with the odor of monkey. It's difficult to describe, but it recalled my experience in Panama, working with spider monkeys and hepatitis. It was so overwhelming it damn near knocked us off our feet. Clearly, the air was recirculating from one room to the next.

The monkey quarters were long, narrow rooms, ten feet wide. In that space were crammed two rows of double-tier cages facing each other, with an aisle of less than an arms-spread width between them.

I didn't like this layout; it was going to put us at much greater risk. These were wild-caught monkeys. They weren't used to being handled and they didn't particularly like humans. They were powerful for their size, quick, and they had long, sharp canine teeth; we could expect them to put up a fight as we tried to work with them.

Nancy and I walked through the outside corridors peering into the rooms and trying not to get too close to either row of cages. These creatures sneeze, cough, and spit. We also didn't want to make eye contact; the males considered that a challenge. It was one of the ways they determined who was dominant in the wild. Even with our quick tour, we could see that most of the animals were animated—excited, angry, curious—while others were still or listless. Their expressions, their general demeanor, reminded me of the patient reports from Zaire 1976.

Fortunately, hygiene was good. It was clear that Hazelton did a good job of taking care of their monkeys. Dalgard had made it his business to contact everyone Hazelton had shipped monkeys to in the past several months and follow up on their health status and condition. Although the physical facilities could have been better, I shuddered to think what we'd be facing if this had happened without a sharp and responsible guy like Dan.

The next morning the team converged at the USAMRIID loading dock at what we military people begrudgingly refer to as "0:Dark:30." We were all dressed in civilian clothes, so as not to attract undue attention. Gene supervised the loading of the equipment and supplies into a couple of unmarked vans. Much of the matériel had been with him at Kitum Cave.

We traveled in an informal convoy off the base, back across the Potomac at the Point of Rocks Bridge, and over toward Route 7, where the Virginia traffic was already maddening. When we got to the monkey house, we parked out back. There had been an article in *The Washington Post* about a possible Ebola transmission at a monkey quarantine facility in Reston. So far, they were either downplaying it or didn't understand its significance, but I knew there would be reporters snooping around and I didn't want to present too much of a target. One reporter dropped a card under the windshield wiper of one of our cars in the parking lot in front, but he didn't stop out back where we were at work.

During his recon visit the day before, Gene Johnson had designated as the staging area a feed storage area directly across the corridor from Room H and with access to the outside. While Gene was in charge of the organization and execution of the containment, he had some powerful help. The USAMRIID aeromedical evacuation team was responsible for pickup and delivery of hazardous virus-infected GIs from anywhere in the world, and they were past masters in dressing up for microbiological disasters. These were the people, even more than the denizens of the hot suites with their blue space suits, who knew the ins and outs of the Racal hoods and orange Tyvac space suits. Mixed in with the teams were experienced primate veterinarians, laboratorians, and animal caretakers with years of BSL 4 work behind them. We'd put on the gear and then step through the door to enter the corridor, which also served as a gray zone where we were disinfected with Clorox spray before we returned to the entry room with the samples.

To be certain nobody had to be sent to the slammer after the day's work, we planned, discussed, and practiced procedures the evening before and the morning of the day we took on Room H: If you get any blood on your suit, wash it off immediately, because if you get a rip or tear and can't see it, you're in trouble; if you get a rip or tear and hot blood leaks in, you're in big trouble. Everybody would be looking out for sharp objects. And then the monkeys—they had literally lived by the law of the jungle and some of them were older, dominant males who had risen to the top of a hierarchy tougher than an inner city gang: these monkeys were fighters. If they grabbed hold of you or sank their canines into you, you'd probably be saying your last farewells to your loved ones through the thick window of the slammer. Even if the particular animal you encountered wasn't positive for Ebola, around 85 percent of captive adult macaques carry herpes B virus, and occasionally have it in their saliva—it's usually fatal in humans and has been responsible for at least twenty-three deaths.

The crew changed into scrub suits in one of the vans out back and came into the building out of view where Captain Elizabeth Hill supervised their donning the containment gear. Liz, in addition to doing a superb job with her crew at Reston, was the officer in charge of the aeromedical evacuation team.

As we entered the hot zone of Room H, some of the monkeys were listless and depressed. Our presence hardly registered with them. But when the healthier and more active ones caught sight of us in the orange suits, they went nuts, banging on the cages, screaming at the top of their lungs, demonstrating aggressive and challenging behaviors.

This was a more dangerous operation than working in your own BSL 4 lab, where there are no traps. In your lab there are no needles lying around, nothing to step in and slip on, no hard or sharp edges that you're not aware of. There's an alarm button to press if you get in trouble. Not so here. This was like going into enemy territory.

Even the suit could be problematic. Although it was lighter in weight and bulk than the hot suite's blue suit, it was still bulky and hampered movement in this narrow and confining work space. Since it was freestanding and not connected to an overhead air supply, you had to contend with a blower unit and several batteries hanging from your belt. The filters, while offering good protection, didn't get rid of that overwhelming monkey smell.

We decided to have two people handle each monkey at the same time. These weren't the "squeeze cages" where you can pull the back of the cage forward and pin the monkey before you jab him. You have to open the cage, and once you do, it's you against the monkey. You do your work quickly, then slam the cage door shut. As difficult as this was for the upper tier of cages, it was even rougher for the lower tier. The cages were near floor level, and you had to squat or crouch down on your knees to get at them. Not an easy task in a Racal suit.

The strategy was to use a mop handle with a U-shaped pad at the end to pin the monkey in the cage, then inject him with ketamine, a general anesthetic. Once that took effect, they'd be injected again with a powerful sedative. If the monkey happened to stick his arm out to grab us, then we'd grab him first while our partner injected him.

Once the animal was out cold, it would be taken out of the cage, bled for a sample, and given the euthanasia drug. When they had died, they'd be opened up, and tissue taken from the liver and spleen, which were dropped into plastic specimen bottles and sealed. The body was then bagged and put in a box. As the day wore on, the stack of boxes grew in the corridor.

Nancy took the first four dead monkeys back with her to Detrick. This time, though, they went in double plastic bags lined with absorbent material, a hard cardboard carton, and a sealed Styrofoam box before being popped into an army van. When she and Ron went into the hot suite and started exploring inside them, the effects of Ebola were obvious. There was spot bleeding and necrosis of the intestines and other organs. These monkeys had been very sick. In fact, from a pathologist's point of view, they looked as if they'd been dead for several days rather than only a few hours. I hated having to kill all those monkeys. But in my mind, there was no question we'd made the right decision. We couldn't let them suffer any more than they had already, and we needed to do whatever we could to assure that their disease remained contained.

In the days after the Room H operation, we were pretty busy with the lab work and analysis. Tom Ksiazek ran ELISAs on the samples, and what Nancy had taken out of the monkeys she analyzed in more detail. Ksiazek was head of rapid diagnosis for USAMRIID, which is considered the military's first line of biological defense. If anything strange comes down out of the sky or wafts in on the breeze or gets coughed or bled from one person to another, Tom was the guy who had to figure it out. He was a lieutenant colonel who had a D.V.M. from Kansas, where he'd gone to school with the Jaaxes, and he had a Ph.D. in epidemiology from Berkeley. Tom is one of the smartest guys I've ever met or worked with in infectious diseases.

It was a trying time for all of us. It was particularly agonizing for Nancy, whose father was dying of cancer back in Kansas. She felt deeply conflicted about where she should be, and every time the phone rang at home, she half expected the worst. During this time we also received another blow, one that brought back a trauma I had been subconsciously suppressing for months. I came into my office one morning and was told by my secretary that Didier Meunier, one of the two French scientists who'd been with Gene Johnson and me on our 1985 Ebola quest into the Central African Republic, had been killed.

It had happened in late July or early August, but the news had just reached us. He was still with the Pasteur Institute, living in Abidjan, the capital of the Ivory Coast. He was planning to spend vacation time in his house in the South of France. His wife and two young boys were already there. Didier was closing up the house in Abidjan, getting ready to leave, when burglars had come in to rob the place. Either out of panic or just plain maliciousness, they shot and killed him. He wasn't even forty yet.

I was very upset. Didier had been a friend. I still have a collection of mounted Central African butterflies that one of his sons gave me. It made me realize how whimsical and arbitrary life was. Just as with disease, we're all running risks all the time. But the talented men and women like Didier Meunier who pass up prestigious and high-paying and comfortable jobs to work in medically, socially, and politically dangerous places are the real heroes of our work. With Reston, we were going to extraordinary lengths to protect people against a potentially lethal threat. And here Didier was killed for no reason, just like that. Sometimes everything seemed so pointless.

We had a meeting with Dan Dalgard at USAMRIID on Sunday evening at seven o'clock that lasted about three hours. It looked like the Ebola was still spreading despite all our efforts. If that turned out to be the case, Dan realized, he was going to have to take some radical step to stop it.

The ELISAs came back the next morning, Monday, December 4. Tom Ksiazek was able to identify Ebola in five out of twenty animals tested from Room H. Tom Geisbert used the electron microscope to confirm the findings. We now knew beyond a shadow of a doubt that, one way or another, Ebola virus was in a second room and therefore might spread further still.

That same morning, Dan Dalgard pulled up in front of the Reston monkey house to find an animal caretaker named Keith out on the grass in his protective jumpsuit. Dan was upset, because he'd given orders that the protective equipment was not to be seen outside the building. But as he got closer, he realized Keith was vomiting and retching onto the curb.

As Dalgard quickly learned, Keith had come to work feeling fine. As he put on his jumpsuit, he began feeling sick; it grew worse and worse until finally, in one of the monkey rooms, he started to lose it. Dalgard called an ambulance and, while they were waiting, had Keith lie down. They didn't have a thermometer that hadn't been used on monkeys, so Dalgard sent one of his associates out to buy one. They took Keith's temperature. It was 101. He felt weak and had no appetite. Keith had had close contact with the quarantined monkeys and now he was sick and febrile. Was this the shoe dropping we'd all been waiting for?

As we now had a human situation, he called CDC. The question was: What do we do with Keith? I thought we ought to put him in the slammer until we knew what he had. But this was Joe McCormick's call, and Joe felt that based on his own experience with Lassa and Ebola in Africa, barrier nursing care would be sufficient to protect everyone and prevent a spread. So despite my offer of our full medical isolation facilities at USAMRIID, which could expand into a thirty-bed unit in a crisis, Joe decided to put him in nearby Fairfax Hospital. Fairfax is an excellent hospital, but no hospital in the country has any significant experience containing rare and exotic hemor-rhagic viruses.

I didn't want to be overly alarmist, because I knew very well that most hoofbeats in our neck of the woods come from horses, but I must admit that when I heard that day that Keith had come down sick, I thought that just maybe we were looking at big trouble. If so, Keith could be the first of many, many victims.

Immediately, I started thinking again about Peter Jahrling and Tom Geisbert. The whiffing incident took place on November 27. Yesterday, December 3, should have been the likely end of their incubation. They'd have to stay healthy now for another week before I'd know for sure if they were out of the woods.

While Keith was being treated at Fairfax Hospital under Joe's supervision, Dan Dalgard walked through Rooms A through D, the four front quarantine rooms at the Reston monkey house. Everything looked physically normal, but there was no sound from the monkey residents. It was eerie. He was very concerned.

Back at USAMRIID, Nancy was looking at slides from monkeys who had died in rooms other than F and H. Histologically, they looked like either SHF or Ebola. Ksiazek had positive antigen-detection ELISAs on a smattering of monkeys from other rooms, provid-ing definite evidence of Ebola transmission. It looked as if Ebola was still spreading throughout the building. Dan began to understand that he was losing control of the situation.

But the next day we got our first piece of good news. Tom Ksiazek reported that Keith's antigen ELISA was negative. When he woke up in the morning, his fever was down and he asked what was for breakfast. Whatever he had, it looked like it wasn't Ebola. We'd dodged a major, major bullet.

But everyone was saying to themselves: *What if?* Dalgard could read the writing on the wall. This was not a virus they could afford to let loose on the general public. He called around to senior Hazelton management and told them he wanted to euthanize all the remaining monkeys and completely decontaminate the building. This was an enormous sacrifice. Dan hated killing monkeys. And every monkey killed represented a loss of up to a thousand bucks. They'd lost close to 200 monkeys so far and this would be another 300 or so.

There were a lot of issues for them to weigh: the financial impact, public health, corporate liability, and public relations. The media still weren't probing very deeply, but they were starting to pick up on the story. In the end, it didn't take long for Dalgard to get the go-ahead.

Dan called to tell me his superiors wanted to turn the entire building and responsibility over to us. That afternoon, the Hazelton lawyers faxed over a document which would give us the whole kit and
caboodle, including all responsibility and liability for the building
and the decon operation. If lightning struck and the building burned,
it was on our heads.

I read it over carefully. My first reaction was: "No fucking way." I
wasn't going to accept full legal responsibility for an Ebola outbreak.
So I went to David Huxsoll.

He liked the training aspect of the mission. We'd never conducted
a full-scale Level 4 operation against an active hot agent in the field
and he thought this would be a good lead-up to any full-fledged
military operation. But he didn't like the idea of the liability any
more than I did. Together we went to Phil Russell.

Russell knew it was something that needed to be done and needed
to be done quickly. "This is in the national interest and we're a
national resource. We gotta get this done. So don't call the fucking
lawyers!"

The fact was, both sides knew this needed to get done and neither
side wanted its lawyers, who are paid to be careful, to gum it up. So I
called Hazelton back and said that we would take responsibility for
our people and our own negligence, but that was all. I faxed them a
letter.

Okay, came back the reply. Let's do it.

In the pre-dawn hours of Tuesday, December 5, we again rendez-
voused at the USAMRIID loading dock and packed our vehicles. We
knew what we were doing this time and we had the absolute A-team:
Gene Johnson, to whom this was like going into Kitum Cave all over
again; Jerry Jaax and some of his best vets; the BSL 4 animal-care
technicians, who'd already looked big-fanged, Ebola-infected monkeys in the face working in the hot suite; and Pierre Rollin, the
brilliant young French epidemiologist who was on sabbatical from
the Pasteur Institute in Paris.

It was a much larger operation than the previous week's. Many
more people were involved. Jerry had to bring along a number of
young veterinarians and animal technicians, many of whom had
never worked in a biohazard suit before. Hell, before last week, Jerry
had never worked in a biohazard suit.

When we got to the monkey house and had staged the vehicles in
back, Gene Johnson gave everyone a final warning. He told them
about Keith and another Hazelton animal handler who'd had a heart

attack several days before. There had already been the human casualties associated with the monkey house, coincidental or not. We still didn't understand the possible scope of airborne transmission, and everyone had to be careful. I gave myself the job of watching over everyone else's shoulder, looking for things they might miss—little things that could have a critical impact.

There were a few reporters snooping around the front of the building, but for some reason that still amazes me, they never wandered around back. Of course, we were careful. If we'd been parading around Reston in our space suits, they would have been on us like a heat rash in August. The few reporters or local citizens I talked to during the course of the operation were just mildly curious, having noticed the trucks' arrival. I told them that we were conducting a standard cleanup operation, and that seemed to satisfy them. We used military-band walkie-talkies to communicate between the rooms and the staging area so we'd be less likely to have our transmissions intercepted.

We were taking out the entire building this time. We considered anything inside potentially threatening and infectious. We changed clothes in the vans. Then everyone went in their scrubs to the staging area, where the aeromedical evacuation team helped them get into their suits and hoods and tape them up. I would not let our people use the lavatories inside, so they'd have to get by in secluded areas behind the pine trees. This would have been bad enough in the summer. But it was a cold and bleak December day.

"If you think this is cold, I was raised on a farm in Kansas—that was cold!" Nancy declared before marching off into the woods.

Even with the cold weather, though, with the air-handling system switched off in the monkey house, it would get pretty hot and sweaty working inside those suits. Everyone was soon walking around with fogged-up faceplates.

Killing hundreds of sentient, intelligent animals is not a pleasant assignment, especially for members of the Veterinary Corps who are dedicated to saving animal life.

Once Jerry and his buddy were suited up with their blowers on, they ventured through the air lock we'd set up and into the building. There they found a situation near chaos. The place had been pretty much deserted for a couple of days as Hazelton figured out what to do, and the monkeys were hungry, thirsty, and pretty crazed. Some were listless, some had runny noses, a few had bloodshot eyes and were sitting above pools of bloody feces. Although they were all going to die, Jerry couldn't tolerate them suffering any more than they had to, so he dispatched some people to the supply rooms to find the monkey biscuits and get them all fed.

Then the operation commenced. A two-person team would go into the room, open a cage, pin the monkey with the pole with the U-shaped pad, then stick the animal with anesthetic and wait for that to take effect. Then the next stick, then carrying it out to the final area, where they took a blood sample before administering the lethal injection. The team members there would wait for the monkey to die, then do a quick and systematic necropsy in which spleen and liver tissue would be snipped out with surgical scissors.

It had taxed the building's limited resources to dispose of the bodies of the monkeys that had died or been euthanized up to this point. Many of the monkeys could be destroyed in an incinerator the company had at another location. But all of the monkeys who looked sick and who we were going to autopsy had to be transported to USAMRIID for ultimate disposal.

This was just the first day of three needed to complete this job—a massive operation. Then we would have to decontaminate the entire building.

Now, dead monkeys can be different things to different people. If they were diagnostic samples, then we would legally and legitimately move them from Point A (Reston) to Point B (Fort Detrick). But if they were medical waste, then we had a big fat problem. We were not authorized to transport medical waste in Virginia or Maryland or across a state line, nor were our vehicles inspected for their suitability. We couldn't bring medical waste onto the base, and we were not authorized to dispose of medical waste even if we happened to bring it in. If the monkeys were considered etiological agents, that was even worse. Then you'd need special Public Health Service permits to pack them up and ship them out. And if we'd been naive or shortsighted enough to declare the dead monkeys all three things, then we'd probably still be there today.

I didn't understand any of this before the fact, but I caught on pretty quick. I got word during the first day of our operation that the base commander had stationed MPs to intercept us if we tried to bring medical waste onto the fort. A tense line of communication developed from me to the USAMRIID executive officer (XO) and from the XO to Dave Huxsoll before going to the commander himself.

After ascertaining the best and worst scenarios, I began rigidly maintaining that these monkeys were nothing like medical waste. We were a research institution and we needed the diagnostic samples. I said to the XO, "I'm going to assume I can bring my samples into my labs. I'm wearing my pager, if there's a problem, you beep me. Otherwise, I'm bringing them back at the end of the day."

The XO didn't like this. He wanted positive voice confirmation, but I said I was too busy and if I didn't hear back . . .

Anyway, we carried our diagnostic sample monkey bodies back with us. We brought what we felt we needed into the hot suite for analysis, then autoclaved the bodies and incinerated them. This is exactly what you do with diagnostic samples or medical waste, by the way, but in government, definition is all-important.

The day had gone well, though, considering the circumstances. No one had gotten stuck, no one had been bitten or clawed by a monkey, Dan Dalgard had diverted a television crew by asking them to get off the property before he had to call authorities, and the McDonald's and the Taco Bell had cleaned up from all the food and soft drinks the soldiers ordered after they'd been sprayed down with Clorox and come out of their suits.

We drove back to Detrick exhausted, unloaded, and staged for the next morning.

At USAMRIID, Ksiazek worked late into the evening in the Level 4 lab testing the day's samples. Several looked like Ebola.

Meanwhile, Peter Jahrling was working in his own Level 4 lab, and Tom Geisbert labored over the electron microscope. Both men were working practically around the clock, trying to isolate the Ebola virus from samples. And two of the samples Peter was using were his and Tom's. He held his breath as he waited to see if the fluorescent antibody test would glow its characteristic green and tell him the Ebola antibodies were reacting to his own blood. But the test came up negative. So far, so good.

The monkey killing took three days. Everyone was extremely careful throughout the operation and the only notable slip-up was one monkey that got out of its cage and ran loose in the building for a while. On the third day it was trapped between the backs of a row of cages and the wall, at which point one of Jerry's men was able to inject it.

Also on the third day, Nancy got word that her father had died. She left USAMRIID, flew out to Kansas, and got there in time for the funeral. When she had first told me about his illness, I had told her to leave the operation to be with him. But Nancy is a military officer who takes her responsibilities very seriously, and she decided that her responsibilities lay here.

Once the euthanasia had been completed, the next order of business was the decon of the entire building. We had to "nuke" it, in the slang of the biohazard team. The way this would be done was to set up about forty electric frying pans on timers and load each pan with paraformaldehyde mixed with water. Heated, this would produce formaldehyde gas, which would get in all the nooks and crannies, and, enhanced by the humidity, kill off everything inside. Before the timer went off, all exterior passages into and out of the building would be sealed with tape.

The remaining food supplies were packed into small bags for incineration. All the lab supplies were incinerated. That left the turds of more than three hundred monkeys. That's a lot of organic material to have to dispose of.

"So what do you want to do with all the monkey shit, Colonel Peters?"

We thought this one out for a while. There was too much to bag and move, not to mention that no one was particularly eager for this assignment. The only solution was to send it all down the sewer. But since it was presumed to be contaminated by one or more serious viruses, it had to be disinfected first. I consulted with some of our experts, and we decided to have it all soaked in Clorox before being dumped.

But wait! Word came from Virginia environmental authorities that you can't put disinfectant down the sewer. That was against the rules.

By this time I'd had it with everybody's rules and regulations. I was trying to run an operation no one had ever run before because I considered it my job, and I was getting pretty tired of all these unnecessary obstacles.

I finally lost it and angrily said, "Okay, guys. If you can't handle putting Clorox into your precious sewer on a one-time basis to deal with a biohazard emergency, fine. But you figure out what to do with the monkey shit, because I've already given you my best ideas and I'm getting ready to go home and leave it for *you*."

There was a short pause. Then came word from those same Virginia environmental authorities: "Okay, do it."

And so we did.

Interestingly enough, it was right around this time that the owner of the monkey house property had contacted Hazelton and expressed fear that with all that was going on, particularly if there did turn out to be an Ebola outbreak, it might adversely affect the value of the property. Yeah, I guess I could see that.

Once again we returned to the routine of our work, with the added job of finishing off the Reston monkey laboratory analysis and following through on new research leads from the outbreak.

Epidemiological studies had shown that all of the Ebola-infected monkeys came from a single supplier outside of Manila. After the devastation at Reston, the supplier claimed they had completely depopulated their colony, and were starting from scratch with a new population of wild monkeys. Shortly after the first of the year, 1990, they began exporting again.

Then later in January I got a nocturnal phone call from Dan Dalgard. Monkeys were dying again in Reston and he was suspicious; he had personally necropsied them and they had enlarged firm spleens, "like a big salami when you cut them." They were in the freezer. He hadn't yet come to grips with whether or not he wanted to know the final answer. He asked me if we could take a look and give him the word privately.

I very much liked and respected Dalgard, but I quickly replied, "No way, Dan. Because if we found anything, one of us would have to call CDC." But he decided to give us samples anyway, and sent over some frozen monkeys.

Unless you've actually done it, it's doubtful you have any idea how long it takes to thaw out a monkey—hours and hours. When we did, though, Ksiazek ran his tests. They were positive for Ebola. Dalgard called CDC.

I couldn't believe it; the whole thing was starting again! And all the infected animals came from the same supplier. This time, nobody asked me what I thought. If they had, I would have said whack them all and scrub the building down again; don't take any chances. After all, the Reston virus looked very much like the Zaire virus—we couldn't tell them apart by our serologic tests or under the electron microscope. There was very little genetic data for filovirus strains and particularly the Reston virus, and even then we wouldn't know what it meant in terms of human virulence.

The counterargument was summed up by George Pucak, a senior veterinarian at Hazelton: simian B twenty-eight, Ebola Reston zero. He was pointing out that over the years the natural herpes virus of macaque monkeys had killed a number of people working with captive animals but that the virus at Reston still hadn't killed anybody. Hazelton wanted to try to ride out the storm and keep infected monkeys in the facility with extra precautions for the caretakers, and the state of Virginia and CDC agreed with them.

As the new epidemic spread through the monkeys in Reston, Dalgard made a very significant observation. Many of the monkeys had different clinical signs: nasal discharge and respiratory signs. Would we send someone over to take samples and rule out another virus, perhaps even measles? I agreed to do so, but I wasn't sending anyone in without a space suit. When we got the samples, no second virus was found. The secretions had enough Ebola in them for Tom Geisbert to find virus directly with the electron microscope. Peter grew out a million infectious units per milliliter of snot, and Tom Ksiazek's test was strongly positive. Nancy's crew found large patches of pneumonia in many of the lungs tested. Some of these findings were present in the last year's epidemic, but now they had assumed such proportions that they rang alarm bells in an experienced clinician like Dalgard.

Dan continued to send us samples to follow the course of the epidemic. Then, around the middle of February, a rather large animal caretaker known as Tiny was doing a necropsy on a dead monkey. While he was dissecting the liver, he happened to slice his thumb with the scalpel. There was a lot of blood.

The liver was sent over to USAMRIID. Tom Geisbert prepared it and examined it under the electron microscope the next day. It was wall-to-wall with Ebola. Tom Ksiazek ran an antigen ELISA. This was definitely an "Oh shit!" moment. Tiny's cut had been deep. This was a major exposure, much greater than anyone else at Reston had taken. It was exactly analogous to Cochabamba, cutting into cadaver flesh. We were afraid this was the one that would do it, that Tiny was a goner.

But he remained alive and well. He didn't even go into the hospital. In fact, he disappeared for a while, which made us all very nervous. We chased him down for several days. It turned out he had gone to another hospital to have his big toe removed because of diabetes. He tolerated the procedure well and did not bleed excessively. But when we got to him and tested his blood, he had very definitely sero-converted. Pierre was able to find traces of virus in Tiny's early blood but only about one millionth of what we would have expected; clearly, this virus was not nearly as dangerous as the Zaire one, at least to humans. What was so different about it? We just didn't know. And to confound things, the CDC crew had run down the other three animal caretakers exposed daily in the Reston facility and they also had antibodies to the virus! There were no symptoms of disease in any of them, nor was there any exposure to the virus such as Tiny had undergone. They were presumably infected by droplets or aerosols in spite of their precautions with surgical masks, protective clothing, and gloves.

The monkeys continued to die; when even one monkey in a quarantine room sickened, the entire room eventually became infected. One particularly scary room was infection-free for over a month and then came down with a single infected animal, followed by a holocaust of dead macaques.

We didn't understand then and don't understand today the basic mechanisms at play. Why didn't the Reston virus lead to the fever and shock that are the rule with the Zaire virus? Why didn't the infected workers bleed internally and from their body orifices? Ebola Zaire killed most of the humans it infected, and so far the Reston virus hasn't even made anyone mildly ill. It's like the apocryphal twin brothers in Texas: they looked just alike but one grew up to be a preacher and the other became a horse thief.

How did the virus spread? I am very suspicious (but can't prove scientifically beyond any shadow of a doubt) that the monkeys' lungs and secretions were filled with virus-containing fluid and that small particles of this fluid were airborne in the monkey rooms and spread to at least one uninvolved room. Was this a mutation and adaptation that occurred in February 1990 and caused the spread? If so, could it happen with the Zaire subtype?

Tony Sanchez has been working on the genetics of the Ebola viruses, and I say "viruses" because his data show that there are four distinct subtypes of Ebola: Zaire, Sudan, Reston, and Côte d'Ivoire. There will be arguments about whether there are four subtypes or really four different viruses and about what genetic differences are responsible for their differing properties. Until we are prepared to answer some of the questions with real data, we have to remain in a dangerous limbo. These are RNA viruses, with their high mutation rates giving them a built-in source of genetic variation that can come into play as they move from their reservoir into humans and then pass from human to human or monkey to monkey.

Maybe Reston could harm people and maybe it couldn't. But it damn sure devastated some close primate cousins of ours. The next critical task for us was to figure out where this particular virus strain had come from, and I asked Curt Hayes, the commander of the navy lab in Manila, who was working closely with the Philippine Research Institute for Tropical Medicine, if I could send Tom Ksiazek over to test monkeys from the various importers. I'd known Curt from way back, and he agreed that this would be a good idea as long as his people were part of the research.

Shortly after this, I got a call from Joe McCormick, who wanted in on the investigation too. He said the people over there might think they were good, but he'd actually dealt with Ebola and they hadn't.

I pointed out to Joe that CDC already had a guy over there—Mark White, who was in the Foreign Epidemiology Training Program. Of course, what I failed to appreciate was that the Epidemic Intelligence Service, while an integral part of CDC, had no real relationship to the Center for Infectious Diseases.

But Mark wasn't an Ebola person, Joe said, pointing out his own substantive fieldwork in both 1976 and 1979.

I remember him saying to me, "You've got to get me over there." "I'm not blocking you," I replied, and agreed to call Curt.

When I spoke to him, I said that I wanted Ksiazek there; anyone else Curt wanted or would accept was fine with us. Curt didn't seem to be sure of the point of having anyone else since he already had Mark White, and Joe ended up going around the U.S. embassy in Manila and appealing directly to the Philippine Health Ministry authorities. The next thing I knew, a few days later Joe was no longer chief of viral Special Pathogens at CDC. He had been replaced on an acting basis by Ken Hermann; Susan Fisher-Hoch remained as his deputy. Sue was a British physician who'd worked on Legionnaires', Lassa, Ebola, and other hemorrhagic fevers. Joe migrated to HIV/AIDS and later to the Malaria Branch of the Division of Parasitic Diseases because of his African experience, and I was told he was planning on leaving CDC for a job elsewhere. (In another of the infectious disease community's seemingly endless personal and personnel reshuffles, Joe and his wife divorced and he married Sue on a skiing holiday in Vail, Colorado, in 1992.)

The Philippines was not a great place to do epidemiological field work and scientific investigation in 1990. Corazon Aquino had recently come to power in place of Ferdinand Marcos in the wake of her husband's assassination, the American presence at the huge Subic Bay naval base was a major political hot button, and political insurgency was rife in Mindanao, the island where the monkeys came from.

With Curt Hayes and Betsy Miranda, Tom Ksiazek evaluated the three major export facilities in Manila licensed by the Philippine government and ran ELISAs on samples from monkeys that had died there. They also took samples from trappers in Mindanao and learned that a major snack there was uncooked marinated monkey meat.

It turned out that one of the three establishments didn't use wild monkeys; they only bred animals, so it was eliminated. The other two, known at the CDC as importer A and B, both bought their animals from trappers. Importer A's death rate among captive monkeys was about 20 percent. Importer B's was about 40. Tom's tests showed no Ebola at importer A and suggested importer B's high mortality rate was due to Ebola. Once again, the people at the importer with the Ebola—the same one who had shipped the monkeys in the Reston outbreak—claimed they were going to depopulate, decontaminate, and start again. We have no way of knowing for sure what happened. At that point, we just didn't have the clout in the Philippines to be able to find out.

The Reston episode had an unsatisfactory, unfinished feeling to it; we never found out where the virus came from. We were gratified by our ability to analyze what was happening and contain it, but none of us felt we'd achieved anything that was going to give us a jump on the problem in the future. If it happened again and our tests showed Ebola Reston, I don't think I'd be confident enough to say, "Okay, this one isn't dangerous to humans; let's not sweat it." As had happened so many times before, Ebola went into hiding.

It broke out again in April 1995, in the town of Kikwit in the southwestern part of Zaire. Kikwit is much larger than Yambuku, with about 400,000 people. But in most ways it is really like a big Yambuku, like a big village. And this time the virus was the deadly Zaire subtype. Out of 318 known cases, the kill rate was 77 percent.

By then, I was CDC's chief of viral Special Pathogens. When the samples came in from Zaire, Tom Ksiazek applied the tests that had been developed and proven during the Reston episode and made the diagnosis in a matter of hours. Tony Sanchez was ready with the genetic analysis of polymerase chain reaction products and within a couple of days we knew we were dealing with a strain of the Zaire subtype virus very similar to the 1976 killer. Then came the question: What should CDC do? We were frankly tired; the hantavirus outbreak had worn morale dangerously low. Money had come, but no new bodies to replace attrition and build up the laboratory. After a long discussion, we decided we had to respond.

After the official request for assistance from the government of Zaire came in, I sent a top-level team, including Pierre Rollin, who had joined me at CDC, and two other highly talented people, Ali Khan and Philippe Calain. Ali had been with the Epidemic Intelligence Service (EIS)—CDC's epidemiology training program—and had distinguished himself in many ways since then. Philippe had completed a postdoctoral fellowship in molecular biology in his home country of Switzerland before continuing at CDC. As soon as they arrived in Kikwit by way of Kinshasa, Ali began mapping the lines of transmission within the area, getting a fix on the outbreak from both a big picture and a local perspective, while Pierre and Philippe helped get the devastated Kikwit General Hospital functioning again.

The international team that poured into Kikwit studied the disease and helped to contain it, but we can't really say that they cured anyone. Patients either got better or they died, and the best the authorities could do was keep others away from them until nature took its course one way or the other. One thing they did confirm was that barrier nursing—special masks designed to prevent tuberculosis spread, gowns, and gloves—did prevent disease spread.

But when the Kikwit outbreak was over, we still didn't know where the reservoir was, we didn't know where the index case had contracted the virus, we didn't understand what made it so brutal and Reston so benign to human beings.

After the January–February 1990 outbreak, which we referred to as Reston II, the monkey house never reopened. It became a derelict curiosity and a symbol of what almost happened and could still happen at any time. Eventually, it was torn down.

The Reston virus has cropped up again from time to time, in Siena, Italy, in 1992, and most recently in Hazelton's quarantine facility in Alice, Texas. The supplier was the same in each case. Each time, it wiped out its simian hosts and left the humans alone. After the 1996 outbreak in Alice, Tom Ksiazek, Chuck Fulhorst, and Ali Khan went back to the Philippines and looked again, trying to find the reservoir. They didn't find it, but I'm not convinced they got the cooperation we need as scientists if we're going to get at the truth. I think political and economic considerations took precedence. As Fred Murphy has been known to remark: "In the monkey business, there's a lot of monkey business."

The entire genome of Ebola Zaire has now been sequenced by Tony Sanchez at CDC and he's very close to doing the same for Ebola Reston, which will tell us what the genetic differences are between the two strains. What it won't tell us is the reason for differences in effect. For that, we have to rely on observation: feet up or feet down.

So what are the lessons? For one thing, it is that bad things can happen and we can't always predict them. We were blindsided by this one. Fortunately, we had people who could deal with it, but I wouldn't take that for granted. If some new virus happened again today, we would not be a lot more prepared, and the next time it could be monkeys again, or horses, or dogs, or people. The nature of viral diversity, replication, and transmission is such that we just won't know. So we have to be ready, and we have to have surveillance around the world to give us the jump on it.

On the bright side, the quarantine system under which the Reston monkey house operated does work in terms of controlling serious and/or mysterious disease brought in from outside the country, particularly when you have people like Dan Dalgard overseeing it. Though the legal requirement was a quarantine period of thirty days, once he realized he had a problem on his hands, he ordered his monkeys held indefinitely, until the problem was sorted out. The quarantine system bought us the time to react. And as a result of the lessons learned at Reston, requirements and procedures were tightened further and the people who took care of the monkeys were all properly equipped, so that by the time of the outbreak in Alice, Texas, what could have been a major public health threat was kept under tight control and there were no significant human exposures.

But many questions remain, both tantalizing and troubling. I ask myself: What if Ebola Reston had been the first filovirus to present itself? Then, when samples came in from Ebola Zaire, would we have said, "Oh, don't worry about that, it's just Ebola. It wipes out monkeys, but it doesn't do much to people"? By the time we realized it was killing people in Africa, it could have spread all through our own building. How would that have looked: the Centers for Disease Control at ground zero for a killer outbreak?

I like to think I still would have said, "We don't know much about this virus; it comes from nowhere. Even though it doesn't do anything to people, look what it does to monkeys. Let's work with it at

Level Three and be very careful until we know more about it." And even at that level, I think we could have had some accidents.

Keep in mind that so far we've had only a handful of human exposures to the Reston strain, so I wouldn't bet the farm that we know all there is to know about its potential for human infectivity and harm.

For some reason which we do not understand, certain patients in the Kikwit epidemic were much more dangerously infective than others. We referred to them as "super spreaders." Why were they apparently so dangerous and what does it tell us about Ebola?

A doctor's most useful instrument is the "retrospectoscope." I think the ultimate lesson of all this is that we have to bring our knowledge up to a certain level so that we can be more competent in exploring threats. We should be developing antiviral drugs and prototype vaccines for filoviruses, so that if a specific threat confronts us, we'll have the knowledge base to go about meeting it. Serious potential problems justify serious research efforts.

We can't just wait for it to happen and then react. I would put it like this: "If you're going to want apples, then you'd better plant the tree." You've got to take the time to water, to nurture, to wait for growth.

I think it's time to start planting those trees.

10

The Gulf War
and Beyond

T HE MISSION of the Army Medical Department has always been the protection of the soldier, and the bulk of that effort has gone into protecting him or her from naturally occurring threats. The protection against chemical and biological warfare agents had been ongoing but largely theoretical until the summer 1990, when Iraq invaded Kuwait.

Saddam Hussein had already been in a protracted eight-year war with Iran, in addition to his domestic aggression against Iraq's own Kurdish population. I had seen medical photographs of the grotesque weeping sores and denuded skin of Iranian soldiers sent to the West for treatment after the Iraqi use of mustard gas, and they are worse than the worst burns I have ever seen in a hospital, with attendant inflammation, extreme ocular sensitivity to light, and horrible scrotal lesions. The suffering and later disability must have been terrible. It is a "good" strategic weapon in that victims require intensive supportive care which would quickly overwhelm any medical infrastructure. Mustard gas has another nasty feature. It is in fact an oily liquid that gives off toxic vapors; on the skin or clothes of casualties, it is readily transferred to their buddies and to medical personnel who touch them. It also remains in the environment for a time.

When the United States, in a coalition with Saudi Arabia and other nations, sent troops to the Middle East, it became clear that if Iraq didn't back down, there was likely to be an extensive ground war in the Persian Gulf region. And given his past performance, there was every reason to believe that Saddam Hussein would use every weapon in his arsenal.

For me, this was more than a theoretical concern. I'd seen the effects of neurotoxins up close. Back when I was a medical resident in Dallas, I was called to the Parkland Hospital emergency room to see a forty-five-year-old man who had arrived in a screaming ambulance because of severe convulsions. We had no clue what was wrong with the patient as we struggled to keep him alive and try to figure it out. He was unresponsive to any stimulus and would convulse frequently for several minutes at a time. Both pupils were pinpoint. He did not seem to have any reflexes that could be determined between convulsions. We were thinking in terms of a stroke, possibly with bleeding into one of the deep primitive areas of the brain, reducing him to a vegetative state with his physiology deeply disturbed.

As we argued with the resident who controlled the ICU, I continued to examine the patient. The resident had a full house and would have to move an MI (myocardial infarction) if we wanted to bring this guy in. He didn't want to do that because he figured our man was a goner and at best would survive as a brain stem preparation in a chronic-care facility or "vegetable farm." The patient, meanwhile, was showing an extremely slow heartbeat, and although we had suctioned out his excess pharyngeal secretions before intubating him, he was pouring out fluid and it was very difficult to give him a proper lung air exchange because we couldn't readily move air out of his lungs. His heartbeat slowed below thirty, very unusual because his blood pressure had gone to shit. I gave him a dose of atropine to speed things up, with little change in his pulse. There seemed little to lose under the circumstances, so I hit him again with the drug and did get some effect this time.

As we took him up on the elevator to the ICU, something from pharmacology class stirred deep in my memory. This man had every evidence of blockage of the enzyme acetylcholinesterase. The chemical acetylcholine is used by many neurons, or nerve cells, to "tell" the next neuron what to do; it's an important bridge that activates the next part of the neural network or makes a muscle contract. After its action, it must be destroyed, and this job is done by acetylcholinester-ase. If the enzyme is blocked, the chemical just stays around and continues to stimulate the neurons.

I had worked with anesthetized dogs in the physiology lab, where we had played through the different scenarios using the many types of clinically important drugs that affect this crucial nervous system chemical transmitter and the counterbalancing enzyme, and had some firsthand experience with the balances and some of the effects. But we had never seen anything this extreme in the dog lab and I had never treated a human teetering on the brink of physiological death.

In the ICU I plunged into the reference books while the intern juggled more atropine to speed up the pulse and keep down the secretions, pressors to raise his blood pressure, suctioning of secre-tions, the settings on the respirator to get enough oxygen into his lungs. This picture fit that of a poisoned enzyme very well and forti-fied us to continue with massive atropine therapy. And over the course of several days, the patient recovered!

It turned out that the patient was an alcoholic living with his sister and her husband. To be sure he had ready access to booze, he kept Thunderbird wine stashed all around the house. His brother-in-law had sprayed his rose bushes with an insecticide and placed the un-used portion in a convenient container he found close at hand—an empty Thunderbird bottle. One evening while searching for part of his stash, the patient came upon the partially filled bottle, guzzled it, and almost ended his life.

Why am I telling this story here? Well, the organophosphate insec-ticides have exactly the same mechanism of action as sarin and other nerve gases. They bind to the active part of the acetylcholinesterase enzyme and form a strong covalent bond with the enzyme, rendering it inactive until more enzyme is synthesized by the body.

Now imagine the aforementioned patient in a battlefield situation, multiplied by thousands. No wonder the military has invested in ways to detect these gases in the atmosphere before the exposure takes place, has developed special suits to protect against them, spends so much time training soldiers in chemical warfare, and has vowed not to be the first antagonist to use it. The fear factor alone, after seeing helplessly convulsing, dying GIs, would have a severe impact on troops.

The treatment for nerve gas poisoning is not well worked out because human studies, obviously, cannot be done. Based on the best possible animal studies, soldiers can be issued automated syringes which provide a dose of atropine and other drugs (try giving yourself an injection if you're not an experienced diabetic), but there is no guarantee what condition they will be in even if they administer treatment to themselves and/or a buddy. Furthermore, we have no idea whether there may be long-term effects in survivors, as has been strongly suggested by some of the experiments with animal brains.

So as far as the army is concerned, this is very, very serious stuff indeed.

There are many ways to conduct chemical and biological warfare. Military leaders don't like them because they don't conform to the traditional means by which the military goes about its business. While I certainly do not advocate chemical or biological warfare—or any warfare, for that matter—I must admit to always having been somewhat perplexed over what was so much more horrible and emotionally repellant about nerve gas or anthrax, say, compared with bullets or bombs. Death in combat is pretty damn horrible no matter how it's carried out. And death in civilian populations is pretty damn emotionally repellent whether it's the result of nerve gas carried in on the wind or bombs falling out of airplanes. The bottom line is that we all have to keep the pressure on the rogue nations that are using war as an instrument of policy or genocide, and we must actively pursue verifiable treaties to stop the proliferation of nuclear, biological, and chemical weapons that result in indiscriminate mass casualties.

I do have a theory, though. I think this is one of those cases where it's the generals, rather than the politicians or the public, who are, and have been, calling the tune. This is not the way generals have been used to fighting wars and so they reject the utility of poison gas and particularly biological agents. This is simply not the way real men fight wars. You set up machine-gun emplacements and you take bridges, without worrying about whether your troops are upwind or downwind from a cloud of germs or gas. The technical complexity and lack of established doctrine for use in combat also has a lot to do with the reluctance among military planners to use such weapons.

The antipathy of the world's generals hasn't stopped major combatants from including such agents in their arsenals, however. Crude attempts at chemical warfare have been used for 4,000 years, beginning with toxic smoke employed during biblical times. Indians, Chinese, Greeks, even the British tried their hands, always with limited success. In modern terms, we would say their delivery systems were not "optimized."

This all changed dramatically on April 22, 1915, when the Germans opened a four-mile gap in the French defenses at Ypres by releasing a cloud of toxic chlorine gas. Five thousand French soldiers died and another 10,000 were injured. This tactic both upped the ante and changed the rules of combat. During the remainder of World War I, both sides liberally used chlorine, phosgene, and mustard gas to kill more than 90,000 people and leave another 1,300,000 wounded. The casualties, particularly by alpha-chloroethyl sulfite—known as mustard gas because of its distinct odor—were so hideous that the 1925 Geneva Convention's "Protocol for the Prohibition of the Use in War of Asphyxiating, Poisonous or Other Gases, and of Bacteriological Methods of Warfare" was readily adopted by most countries of the world, although the U.S. Senate ultimately failed to approve it.

Biological warfare can be reasonably traced back to the fourteenth century when the Golden Horde of Genghis Khan came west. Plague, or the Black Death, is caused by the bacterium *Yersinia pestis*. It was established on the Mongolian steppes among burrowing wild rodents and spread from rodent to rodent by flea bites. Khan's armies inadvertently brought infected fleas west with them on one of the most dangerous rodent reservoirs, the black rat or *Rattus rattus,* which lived close to humans. As a result, they converted the plague from a rural disease to one that could strike cities.

The episode in question occurred in 1346 in the Crimean city of Kaffa, now called Feodosiya. The Tartars surrounded the city on the coast of the Black Sea and laid siege, but to no avail. Then plague appeared among their army; epidemic disease is a common accompaniment of crowds of people with poor hygiene, particularly in military encampments. They decided to deal with the besieged settlement by catapulting plague-ridden cadavers over the walls. The result was an epidemic that eventually led to the abandonment of the port and may have played a role in hastening the transport of plague to Italy as the surviving inhabitants fled.

In June 1763, the British forces under the command of Jeffrey Amherst were reported to have given blankets from smallpox patients to the Indians. There are also stories of this tactic being used by the American cavalry during the Indian Wars, one of the many factors leading to distrust of government authorities during the Four Corners hantavirus outbreak in 1993.

Even after the Geneva Protocol, research and development efforts continued. You never knew when the "other guy" would break the treaty and you wanted to be prepared. Italy gassed totally unprepared Ethiopians in 1935. We know for a fact that Germany had an interest in both chemical and biological weapons in World War II and did a lot of work on nerve agents. The new German nerve agents such as tabun, developed under the leadership of Dr. Gerhard Schrader, resulted in paralysis and central nervous system toxicity, and often death. This is what we do with common pesticides—that is, wage chemical warfare against household and agricultural pests. And that is why my patient at Parkland who downed the contents of the Thunderbird bottle was such a good example of a chemical agent casualty. In fact, Dr. Schrader was working on a new insecticide in 1936 when he came up with tabun. Two years later, he developed a compound related to tabun that was ten times more toxic. Schrader called this new nerve agent sarin, after the four key people associated with its development: himself, Ambros, Rudriger, and van der Linde.

In spite of the invention of nerve gases, chemical weapons were scarcely used by any of the combatants during World War II, with one terrible exception: 30 tons of Zyklon B, a hydrogen cyanide-based insecticide, were used to kill millions in the concentration camps.

U.S. chemical warfare defenses were not part of the day-to-day training of troops until intelligence information in the early 1980s showed the Warsaw Pact troops busily practicing during chemical warfare maneuvers. This prompted a spurt of activity in chemical but not biological warfare defense. The protective suits, newer masks, procedures to decontaminate troops, field detectors, and nerve gas antidotes were all developed or improved and actually deployed as part of practical training.

In contrast, World War II saw active development and field testing of biological weapons by Japan with Dr. Shiro Ishii's notorious Unit 731. Ishii operated out of a prison camp, Pingfan in Manchuria, beginning in the late 1930s, using Chinese and eventually Allied prisoners in unimaginably cruel experiments. Like most human experimentation conducted by sociopaths in the name of scientific research, his methodology and data were so flawed as to be all but useless. These sadists are more interested in the experience than the experiment. Shamefully, in my opinion, as we did with Wernher von Braun and the German rocket program at the end of the war, we pardoned Ishii and his crew and tried to hide their crimes so we could obtain the fruit of their labors. We didn't get very much in the bargain.

In our own country, some very bright American scientists signed on at what was then known as Camp Detrick, at the foot of the Catoctin Mountains of western Maryland. A large and formidable fence was erected around their compound, and a very high clearance was needed to get "behind the fence." Many of the workers, particularly the lower-echelon ones, didn't even know what they were dealing with. They might not know, for example, that "Agent X" was, in fact, anthrax.

But it wasn't just workers. Unsuspecting citizens also became subjects. For example, simulated anthrax spores—*Bacillus subtilis* spores—were released into the New York City subway system, and *Serratia marcescens* bacteria were generated from a boat off the California coast and recovered inland to measure the airborne concentrations over distance.

While these experiments seem reprehensible today, concepts of informed consent and scientific knowledge of the pathogenic potential of *Serratia* bacteria for weakened hosts were not well developed in the 1960s. At the time, the army felt these experiments were crucial in showing the potential of this type of warfare if pathogenic organisms such as tularemia or *Bacillus anthracis* had been used.

There are several strategic uses of chemical and biological warfare. One is to kill as many enemy soldiers as you can—the microbiological equivalent of bullets or bombs. The second is to disable your enemy for a given period of time—to make them just sick enough that they can't fight while you achieve your defined military objective. Sometimes, killing a bunch of your enemy isn't nearly as effective as wounding or disabling them. In Vietnam-style ground combat, for instance, if you shot and killed an American soldier, you'd have taken out one fighting man. If you severely wounded that same person, however, you'd also have taken out the two, three, four, or five men necessary to treat and medevac him. If the disease is then contagious, your initial "hit" can be multiple.

In strategic planning, there was an initial presumption that all biological warfare would be aerosol-delivered, downwind, straight into the lungs of the enemy. Infectious agents have a great advantage over biological or chemical toxins, since you don't have to deliver the full dose. Once it reaches its target, it will grow on its own to a fully effective level.

The developmental experimentation, then, becomes pretty

straightforward. You calculate the minimal infectious dose. You figure out how much of the agent you can grow given your time and space considerations and how well you can concentrate it. You experiment to see how stable it is in aerosol form. And you have your physicists develop equations plotting the downwind dispersal patterns.

Of course, utility is limited by such factors as atmospheric conditions and the enemy's protective measures, but by the same token, you can say that the efficiency of a bullet is limited by the protection of a tank, and the effectiveness of a tank is limited by conditions of terrain. The bottom line remains the same: it's one more weapon of mass destruction in the arsenal.

There have been attempts to weaponize many diseases. Plague, tularemia, VEE, anthrax, Q fever, Junin, and Machupo have often been mentioned as aerosol infectious agents to worry about. Rift Valley fever, for example, has pretty devastating effects, and if mosquitoes were present, it might keep going either in sheep or in cows. The most formidable kind of biological agent is one that spreads from animal to animal or person to person, which is why aerosol-transmitted agents such as foot-and-mouth disease for cattle, or smallpox for humans, are so scary.

Smallpox would be a horrific biological warfare agent. It can be disseminated in aerosol, it spreads rapidly from person to person, and since its eradication in 1980, there has been very little vaccination in the civilian population. If vaccine stocks were still usable, mass vaccination would be time-consuming, and immune globulin from the blood of vaccinees, used to treat vaccine reactions, would be in very short supply.

It just so happened that USAMRIID was between commanders. Dave Huxsoll had left, and Ron Williams, his designated successor, had not yet arrived to take his place. Charlie Bailey, whose background was in entomology and who had excellent command and leadership skills, was filling the void as acting commander. He told me that we had to get on board Desert Shield, as the operation had been named, quickly.

The problem was, strategic issues were being handled at a Top Secret or Code Word level, and I and my people were cleared only through Secret. Charlie was the only one at USAMRIID with Top Secret clearance. In the wake of the Walker family spy case against the navy in 1985, security clearances had been cut way back. To complicate the matter even further, funding cutbacks meant that all clearances were taking much longer to get through the system.

This is not just a bureaucratic hurdle. The military intelligence people are justifiably quite serious about these designations. This point was driven home to me once when I was giving a briefing to other medical personnel with Secret clearances similar to mine. Before I began, the security officer asked me if I was going to use a microphone. I told him I hadn't thought about it yet, that it would depend on how many people there were in the room and how well they said they could hear me. And anyway, what difference did it make to him?

Well, it turned out it made a lot of difference to him, because his people had determined that if I spoke in a normal voice, the sound would not reach beyond the inside of the auditorium. But if my voice was going to be amplified with a microphone, the sound could carry out into the halls and they would then have to post guards at strategic points to make certain that no one without a Secret clearance could get close enough to pick up any of what I would be saying. By the same token, all notes taken by participants at the meeting would have to be turned in and sent by the security officer to their own units' security officers for delivery directly to their offices. They could not be entrusted to carry these notes in the outside world on their own.

When it became clear that we couldn't very well help in the Desert Shield effort to defend against potential chemical and biological warfare threats if we didn't know what was going on, the intelligence authorities gave us the temporary "Get Out of Jail Free" cards we'd need to be briefed on Top Secret and Code Word matters. Code Word clearance is part of the SCI—Special Compartmentalized Intelligence—system whereby even with a high security clearance, some types of information are given out only on a strict need-to-know basis.

As soon as the new clearance was in effect, Charlie Bailey, Phil Russell, and I were briefed by the analysts from the tri-service Armed Forces Medical Intelligence Center. AFMIC, located at Fort Detrick, is just what it sounds like—the military analytical center for medical threats against the military, be they naturally occurring or manmade. Since the army is the designated lead agency in defending against military chemical and biological warfare, it was up to us to figure out, based on AFMIC's information, what needed to be done.

The chief concern the AFMIC people expressed was over anthrax.

Because of security considerations, I still can't go into what the evidence was, but suffice it to say, it got everyone's immediate attention.

Anthrax, a bacterium known to science as *Bacillus anthracis,* has probably been a scourge since the beginnings of civilization. Its most notable characteristic is its ability to form spores which are metabolically slowed virtually to zero. They are extremely hardy, protected by a very resistant shell that is somewhat akin to a microscopic acorn, and they can survive in the soil or other places for many years, to be eventually picked up by animals, who die and replenish the spores and may also give the disease to humans.

But here's the kicker: the bacillus itself is not particularly dangerous to encounter as it's growing in its vegetative form, but once it begins liberating toxins, this is what kills you. When the host dies, the bacteria then have the ability to go back into a quiescent phase in the spores, waiting for their next encounter with a convenient host.

With modern hygiene and controls, anthrax isn't a huge problem in nature any longer, but there are three types to which humans are susceptible, and this forms its basis as a weapon.

The first type is cutaneous anthrax, in which the bacterium enters the body through a tiny breach in the skin. The toxin can kill a small area of tissue—you will see a black necrotic area—but this is not life-threatening and it won't make you very sick.

The second type is commonly known as woolsorter's disease, because it was identified in those carding and preparing wool in New England textile mills. Remember, the spores are practically indestructible and they would transfer from the soil to the bodies of sheep and would carry over in their wool. It is also called respiratory anthrax because the victim inhales the spores deep into the lungs, where macrophages deposit them in lymph nodes in the chest. As the spores germinate, the lymph nodes swell and push aside the breathing tubes in the chest while actively growing bacilli flood the bloodstream with toxins, leading to death within two to three days in most of those infected. The right antibiotic can be effective if given early; if given too late, while it sterilizes the bloodstream, too much of the toxin has been released and the patient dies anyway. When I was an intern, we used to refer to this as a "Harvard Death": you kept pumping meds until all the cultures were negative and test results were normal but the patient still succumbed.

The third type is gastrointestinal, in which the patient ingests contaminated meat and develops anthrax with toxemia and a high mortality rate.

Of the three, respiratory anthrax is the one that can be most effectively made into a weapon. It's easy to grow in large quantities and the spores are very stable in aerosol. Before the American offensive program was ended in 1968, the army used to grow a lot of it at Detrick. Building 540, the red-brick tower that served as the army's anthrax pilot plant, was outfitted specifically for this purpose. In spite of rumors to the contrary, *B. anthracis* spores are killed by proper decontamination procedures and the building is now being dismantled.

I've always been fascinated by ruins of past civilizations. I went through Building 540 not long ago and I felt as if I were looking at the ruins of Cold War America. It's dark and quiet now, a rotting hulk with broken furniture and lab equipment and debris scattered about the floors. But at one time there was a centrifuge in there, so large and powerful that before it was turned on, its solid-steel lid had to be bolted down. And the building still houses four 250-gallon fermenters, each capable of cooking up enough spores to infect hundreds of thousands.

At the time of the invasion of Kuwait, we hadn't produced anthrax for a quarter of a century. The crucial question was whether we still knew enough to defend against it in a real-world, real-war situation.

The other big concern of the medical intelligence people was botulism. *Clostridium botulinum* is another bacterium that produces toxin, generally in confined areas where there is not much air present, such as a closed container of food. It represents the most serious form of food poisoning, often leading to death. It causes progressive neuromuscular paralysis, which, without medical intervention, leads to respiratory paralysis. In a battlefield situation, treatment would be excruciatingly difficult. How many soldiers could be ventilated with either a respirator or a mechanical bag at one time? And who'd be left standing to minister to the casualties?

Botulism can't be spread over huge areas, which makes it most useful against a specific target or as a terrorist weapon. But anthrax is capable of very wide coverage—its effects can be felt downwind several miles away. For example, 100 kilograms of anthrax spores could be delivered some still, dark night by an airplane passing upwind of Washngton, D.C., and take out one to three million legislators, generals, journalists, housewives, and other civilians. And we judged that both agents were well within the Iraqi capability to produce effectively and in sufficient quantity to endanger our troops.

At our emergency strategic meeting there were a lot of puckered sphincters. "Saddam Hussein never met a weapon he didn't like," the AFMIC analysts told us. "Tell us what USAMRIID can do."

There were several initial considerations for wind-borne dissemination of CBW agents. Night is the best time. Not only do you have the cover of darkness, but there is no ultraviolet light to damage the aerosol. UV light breaks down viruses and bacteria; in fact, we use it as one of our hot-suite safety measures. At night, you are also more likely to have an air inversion situation where there won't be thermals to lift the killing cloud off the ground. Wind currents have to be favorable, and in winter the prevailing winds from the Gulf blow across the Arabian peninsula, raising the possibility of an attack from a boat with an aerosol generator, similar to the test runs with *Serratia marcescens* in San Francisco decades ago. We also knew that the Scud and military shells could have been fitted with biological warfare warheads; the cluster bombs later reported in the media would have been devastating if outfitted with such warheads.

We needed to know immediately how to counter their chemical and biological warfare capability. We were sending thousands of troops over every day and could go to war at any time.

It took the rest of the day to get my people the security clearance they needed. In the meantime, I had to get myself up to speed. I'm a virologist, but the two most immediate threats were nonvirological. I went through a steep learning curve reading old reports and attending conferences with the bacteriologists at the Institute, the people who really knew these diseases.

With the exception of smallpox, which, absent immunization of target victims, is a devastating biological weapon, the military had never taken a definitive step toward immunizations or other biological warfare defenses. There were a lot of reasons for this. For one thing, there is always a statistical risk attached to any large-scale inoculation program, as the organizers of the swine flu program in 1976 learned so well. Also, even if it doesn't cause any serious disease, it will cause a lot of low-grade reactions which can make a person feel unwell for a few days. When you're trying to gauge useful man-hours in a large fighting force, these are significant considerations.

There are also huge risk-benefit considerations. For example, should we continue to give all the troops smallpox vaccination, with its attendant statistical risk, when smallpox has been eradicated? Dave Huxsoll, our former commander, had spent years trying to get these issues addressed, but until the fat was in the fire, so to speak, it was never the brass' top priority. Paying attention to his concerns and warnings would not only have cost money, it would have necessitated a change in the military mind-set. No one in the Health Services Command believed biological agents would ever be used. And this attitude, of course, percolated all the way up and down the chain of command. The complexity of explaining infectious disease concepts mitigates against dealing with them. No one wants to hear about case-to-infection ratios, spectrum of clinical disease, and how you determine antibiotic effectiveness. Until Saddam, every armor commander would much rather have had another tank than an anthrax vaccine.

We had to start by imparting some very basic understanding to people in the command structure: anthrax is a bacillus. It is a Grampositive bacillus. It forms spores. The bacillus is not particularly dangerous but the spores are highly infectious by aerosol. An aerosol of these spores can be inhaled into the lungs and disable and kill vast numbers of troops. That's why you have to be concerned. That's why we have to talk about protective measures.

As anthrax grows in the system and begins releasing its toxin, it liberates a carrier molecule known as the "protective antigen," which is the key to the toxin attaching to cells in the body. It is a kind of Trojan horse that carries a "lethal factor" molecule or an "edema factor" molecule which actually does the damage. For safety's sake, the vaccine for anthrax was based on immunizing only against the first carrier molecule. If that was neutralized, then the rest of the disease process could be halted. To make the vaccine even safer, it was further inactivated with formalin, which gives it quite a kick when injected into the shoulder. A series of six such injections had been shown to be effective during a trial in New England textile mills, but since then very little additional work had been done. It had never been determined that you really needed all six of these shots. Was three enough? What about two? Could you shorten the schedule? Could you correlate some measure of serum antibodies with protection? We had answered these questions for Rift Valley fever, but they had never been addressed for anthrax. Also, the vaccine had never been stockpiled in large quantity and no pharmaceutical company was set up to produce it. No one wanted to, because the potential number of doses, even though large, was limited; and because of the spore problem, the FDA wouldn't allow its production in a building in which any other drugs or biologicals would ever be produced, meaning a pharmaceutical company needed a dedicated facility.

Whatever vaccine existed had been manufactured for us by the Michigan Department of Health, which did not have the resources for a massive production gear-up. RIID was in the middle of the anthrax vaccine production problem because we had the expertise to advise the responsible parties. But the fact was that this was a licensed vaccine, and it wasn't our job to procure or stockpile licensed vaccines; ultimately the whole problem came to roost elsewhere—Huxsoll had spent an enormous effort to get others to listen.

So the next question we had to wrestle with was: Who would get the vaccine? And before we could determine that, we had to determine whom anthrax was most likely to be used against. Was it the front-line combat troops, or would the Iraqis attempt to launch against support troops preparing the planes and vehicles?

The Surgeon General (of the Army, not the Public Health Service) ultimately recommended vaccination of the Special Forces troops immediately, and everyone else in order of likely exposure as more units of vaccine became available.

USAMRIID staff met regularly with Joe Ledford, the Army Surgeon General, to discuss our concerns. There was no doubt in anyone's mind we could knock the shit out of Saddam's forces in conventional combat, but the chemical and biological threat could become a dangerous equalizer. Saddam's strategy was based on his assumption that the United States would be unwilling to take large casualties. He might believe that if our troops started dropping in large numbers, especially with horrible symptoms, we would just turn around and head home.

Botulism was more problematic. It was a more difficult vaccine to produce, it wasn't available in quantity, and even after administration, it takes several months to develop immunity. Worse, like Ebola and dengue, there are multiple known botulinum types. When the vaccine was developed, only five botulism serotypes had been identified, and we knew it was very unlikely that it protected against the two most recently discovered.

The botulism vaccine was not licensed by the FDA; it was still considered experimental. We wrestled with the issue of compelling a soldier to take a vaccine that could potentially be dangerous. But if the Surgeon General thinks it will preserve the effectiveness of troops in battle he could do so. We really only had enough of it to cover the Special Forces anyway—giving it to anyone else became academic.

Our next problem was even grimmer. Unlike a lot of other armies that bury soldiers where they fall, we try to bring back our dead. The

Armed Forces Institute of Pathology was the group charged with positive identification of returning casualties at Dover Air Force Base in Delaware. What would happen, they needed to know, if corpses came back to Dover full of anthrax or mustard gas or, even worse, with both? We advised them to do the same thing we did in our fieldwork with hemorrhagic diseases: dump in enough Clorox to kill off any remaining bugs, only use the higher concentrations needed to reliably neutralize chemical agents. I'm not sure what the planners' worst-case scenario looked like. I do know that they ordered 10,000 body bags.

All the defenses against biological warfare are not necessarily biological themselves. Beyond the realm of vaccines and immunoglobulin, there are also basic barrier defenses to be considered, such as gas masks and MOPP suits. MOPP stands for Mission-Oriented Protective Posture and is broken down into four stages of defensive response. A MOPP 4 suit is so heavy and bulky that it's virtually impossible to do anything other than stand in it. And if you factor in the oppressive desert heat, it's really not a very viable solution.

A gas mask is adequate protection against most, if not all, biological warfare agents. Though skin contact is a concern, it's not the major risk that aerosol inhalation is. The problem with a gas mask is that you can't wear it all the time, so unless you have some kind of early-warning system, it's not going to be effective either. The Chemical Corps did have machines designed to detect various bugs in the environment, sample air, capture aerosols, and test them. The problem was, they didn't work for biological agents. There were too many false positives to rely on them. If an attack was signaled or a reliable intelligence report was provided, fine—a soldier would know to put on his gas mask before the agent reached him. But unless he already had it on, he could inhale a whole lot of deadly anthrax spores without anyone being the wiser.

Antibiotics will treat and cure anthrax if the medical team gets to it in time—before the bacteria generate a lethal toxemia. That means we would have to know about the attack in time, such as by finding the vehicle with an aerosol generator, or we would get our first clues from casualties (i.e., GIs as canaries). This isn't a perfect defense, but it probably represented our most solid shot.

So which antibiotic should we use?

Whenever a bacterium is sensitive to penicillin, penicillin is always the drug of choice. It is cheap, reliable, and generates few unwanted reactions. For those allergic to penicillin, tetracycline is nearly as effective. Most naturally occurring strains of anthrax are sensitive to penicillin. But if someone is capable of fermenting fifty gallons of anthrax at a throw, might they also be capable of making it resistant to penicillin and tetracycline?

I talked to the anthrax guys about how difficult it is to make a penicillin-resistant anthrax. The answer was: not that difficult.

It became more and more likely that Iraq did, indeed, have the anthrax capability, whether or not they'd been able to develop antibiotic-resistant strains. We learned that they had been buying large-volume fermenters, continuous-flow centrifuges, and seed cultures of both botulism and anthrax. There was no legitimate reason we could think of to import these cultures unless they wanted to weaponize them.

I gave the information from our anthrax group to my contacts in the military intelligence community to get a feeling for the scientific competence of the Iraqis. They all agreed that Iraq had the capability to develop antibiotic-resistant anthrax.

This was a major pucker-factor decision. We had to make a recommendation to the Pentagon on which antibiotic to stockpile and bring into theater in massive quantities. And if we were wrong, it would be the equivalent of supplying hundreds of thousands of American soldiers with defective gas masks.

We analyzed every antibiotic available and approved by the FDA, going back through all the research data and animal models before considering our own studies. After some sober reflection, the drug we finally decided on was ciprofloxacin hydrochloride, manufactured by Miles, Inc. under the brand name Cipro.

Cipro seemed to fit the bill in nearly all respects. It was generally well tolerated and relatively nontoxic. It was administered orally rather than by injection. It was effective over a broad spectrum of bacteria. The disadvantages—that it was new and very expensive—actually worked in our favor. No one was thrilled by how much it was going to cost to supply many thousands of doses "just in case," but because it was so new, proprietary, and still under patent not enough was known yet about its properties to be reliably incorporated into anyone's offensive biological program.

We got together with the anthrax people and quickly worked up a Level 4 experimental protocol in which monkeys were exposed to anthrax by aerosol and then treated with Cipro. The drug worked. The monkeys didn't get sick or release toxins.

In any of these situations, it's vital to have an institutional memory. Bill Tigertt, a retired general and former director of USAMRIID, recalled an experiment they'd done some time ago to determine how long anthrax spores remain viable in the soil. The answer: years, at least. Are those spores susceptible to penicillin? No. What's more, the spores don't all germinate in the body at the same time, so if you stop giving penicillin to monkeys after a couple of weeks, before all the spores have germinated, you're in big trouble.

Tigertt thought this was a real problem, so we went back and modeled it again in monkeys. Monkeys have the same disease pattern for anthrax that humans do, so we were pretty confident our data would be relevant. What we found was that in every case, all spores had germinated within six weeks. So in the case of a possible anthrax exposure, a six-week regimen of Cipro should prevent generation of the deadly toxin. We therefore concluded that a two-week supply of the drug would be sufficient to have on hand in theater, and the rest could be procured and flown over before the initial supply ran out.

We worked out botulism defenses at the same time. We had a fair amount of botulinum immune sera, or globulin, from people immunized at the U.S. BW manufacturing plant at Pine Bluff, Arkansas. But there had been virtually no testing data on humans and we didn't know for sure if it would work on airborne botulism. And even if it was effective, we didn't know how much of it a person would need for protection.

We set up another set of monkey aerosol exposures with a huge variety of permutations in the amount of exposure and dose of globulin.

There were many possibilities for delivery systems. A simple aerosol generator on the back of a jeep could do the trick. More sophisticated methods included putting the agent in a cluster bomb or Scud missile. That scared the shit out of us. The greater military establishment just hadn't taken biological weapons seriously until Desert Shield and now we were paying the price.

"That's the way I'd do it," Bill Patrick said.

We were lucky Bill agreed to come out of retirement to help us out. Had he not been there, we would have been doing a lot more groping in the dark. The simple fact of the matter is that you have to understand the offense to plan a defense and Bill was one of the few people left who actually knew the nuts and bolts of an offensive program.

One of the things the strategic people wanted us to do was figure out how much BW material the Iraqis might have on hand for their bombs and sprayers, as well as estimate what might happen if U.S.

strikes hit bunkers containing BW agents. We calculated the amount of material that could have been produced since the purchase of the large fermenters, assuming they were running full-time at maximum capacity—it was enough to take out Baghdad and maybe even reach Iran if the wind was wrong.

Again, Bill Patrick was the voice of reason and logic. He knew the reality of running such a program, and he'd made a pretty good evaluation of the personnel they had to work with. He felt they had considerably less material. Furthermore, the likelihood that most of it would become airborne after a bomb was dropped on it was nil—he had spent years behind the fence watching and participating in intensive work to try to optimize dissemination of BW agents and had some pretty definite ideas. They could still inflict a fair number of casualties, but there was no way BW material would go beyond the border of Iraq, and it probably wouldn't even be a problem in Baghdad.

Still, there were major strategic considerations to all this, particularly regarding target planning. Military intelligence knew there was a munitions factory in Salman Pak, a densely populated area about fifteen kilometers from Baghdad. If they were producing biological warfare agents there and we hit it with one of our smart bombs, how many kids would get infected with anthrax? Plans were made to minimize the aerosolization of any stored BW munitions.

We were concerned about a whole range of other agents, but there wasn't much we could do about them, which made us very nervous. We had no clue how to defend against genetically manipulated weapons. We didn't think the Iraqis were sophisticated enough to have developed any, but there was intelligence information that the Soviets had been working on such a project in the days before the fall of Communism and it was possible that some of that work could have filtered into Iraq.

They also might have weaponized something like Machupo, but there wasn't enough ribavirin in the entire world to handle the casualties we could have taken. We have essentially no stockpile.

In the end, the Iraqis never unleashed chemical or biological weapons in the Gulf War, and those of us involved in strategic planning had to ask ourselves why not. If they were going to use them at all, one line of reasoning would have it, they should have used them early, when they could have delivered a powerful demoralizing blow to the allies and made them take heavy casualties. Another line of reasoning was that they were being saved as a last resort, when all else failed and they had nothing to lose.

But, in fact, they did have something to lose. Saddam knew this, and I think therein lies the answer. I can't say with any degree of certainty, but my personal guess is that when my fellow Texan Secretary of State James Baker sat down with Iraqi Foreign Minister Tariq Assiz in the fall, he said something to the effect of "Listen, you guys use chemical or biological weapons and we'll turn your fucking capital into a molten, silica billiard table with nuclear weapons and there won't be another Iraqi man, woman, or child living there for the rest of time."

Something like that, at least, was terminology Saddam could clearly understand. Because he was certainly a cold, calculating son of a bitch and one way or another, regardless of how many of his own people he sacrificed, he undoubtedly figured he'd survive this, as, in fact, he did. He must have concluded he'd be better off if he had a country left to rule.

Nonetheless, we didn't get off scot-free, and good people from the United States and the coalition suffered and died for our victory. Nor did most of the Iraqi troops that paid the big price for Saddam's ambitions really understand the game or the stakes. More urgently, what do we know about the Gulf War Syndrome? I certainly don't have any unique insights, but we have to bear in mind the troops' immersion in a foreign environment, the massive pollution from oil field fires, all the possibilities related to the ill-understood health effects of the machines of modern war. I remain unconvinced that we have the data to fully appreciate all the possibilities, and even the definition of a "Gulf War Syndrome" remains difficult.

As we look at the problem, we should keep in mind that there have always been casualties from elements beyond the actual weapons used by the enemy. These are officially called "DNBC," or "disease and nonbattle casualties," and are usually made up of jeep accidents, hepatitis, malaria, and the like. Among these is post-traumatic stress syndrome, which has been a major aftereffect of all modern wars. Jonathan Shay has pointed out that there are remarkable resemblances to such behaviors going back as far as the Greek armies in *The Iliad*. In the Spanish-American War, one of the major DNBC categories was yellow fever. In Vietnam, malaria and virus diseases provided major challenges; later we found the deleterious effects of Agent Orange to be important, though less obvious and much harder to come to grips with. The Gulf War Syndrome may be something much more subtle and even caused by multiple factors. I wish I knew.

But I do know that we as a society owe it to these brave men and women who went to the Gulf to take their problems seriously and address them thoroughly and openly, using the same objective methods of epidemiology and pathogenesis that we employ with acute infectious diseases.

That is only one of the lessons we learned in the Gulf. The biggest lesson, I feel, is that we've got to be prepared for war, not only on our own terms but also on the potential enemy's terms.

We talked some about the difficulty of handling multiple cases of chemical or biological effects. What would happen if either one was combined with radiation injury? The treatment problems become uncontrollable and mortality rates shoot through the ceiling. There are tremendous risks to caregivers, particularly with chemical agents. Before you can even treat a victim, you have to decontaminate him first.

When a civilian hospital emergency room has to deal with a major disaster like a train derailment or multiple car wreck, twenty or thirty critical cases at a time just about stresses their resources to the maximum. In the Gulf War, the twin navy hospital ships *Comfort* and *Mercy*, with full staffs of some of the most talented trauma specialists and support staff in the world, were planning and drilling for months to be able to handle around five hundred critical-care cases a day each. Try to imagine how you would deal with five or ten thousand at a crack. It just can't be done. If there is a chemical attack, by the time we even find out about it, there could be thousands convulsing in the streets and dying on the ground. The triage process itself—deciding who gets treated and in what order—would cause total social disruption.

We have to plan, we have to prepare, and we have to prevent. I don't want to be an alarmist, but I've seen too much and I know too much to be able to bury my head in the sand on this issue.

Nowhere is this more crucial than in the area of terrorism.

The first question we might ask ourselves is why haven't these agents been used more in terrorism in modern times?

I think the key reason is that BW and to a lesser extent CW remain a new frontier, not yet assimilated by terrorist groups, but I don't take too much comfort in that. It is only a matter of time, and that time is running out.

During the Gulf War, Bill Patrick and I had a meeting with the brass at the Pentagon. Not wanting to brave either the traffic or the

winter trek from a far-distant parking lot, we took the Metro from
Rockville. I remember coming up out of the subway station underneath the building, into the vast basement-level shopping and support concourse, and Bill saying, "Oh, what a wonderful place for an
anthrax spore attack!"

Bill was not speaking as some ghoul or weirdo, but as a past
master at a profession that we have forsaken as a nation, and with
good reason. But in spite of our intentions, others don't have the
same ideas about tomorrow's world, and Bill's assessment of our
vulnerability remains with me today. We'd better do something
about it.

For terrorists to make effective chemical and biological agents,
they have to be able to put them together, and they need the expertise
to deliver and use them. That's the basic equation. And with the
dissolution of the Warsaw Pact, among other developments, there are
plenty of people with biological warfare experience on the market.

Take a scientist in Minsk who loses his job in Belarus because he's
a Russian national. He's pissed off, he still has good contacts in the
Soviet establishment, and he has access to whatever he wants to sell
to the highest bidder, who, by the way, is going to pay him more
money than he's ever made in his life. One of the "nice" things about
chemical or biological weaponry is that you don't have to steal and
transport a kilogram of plutonium. All you need is a teeny-tiny seed
of whatever the microorganism is, which later will be grown to 10^7
or 10^8 per milliliter.

This is a very difficult concept to get into the heads of people
attuned to things like plutonium that don't replicate. It would be no
big trick, for instance, to take something like tick-borne encephalitis,
which grows to high titer in mice, inoculate the mice, wait for their
illness to tell you their brain was full of virus, and then go on to the
next set of mice. It's Malthus' old equation with a vengeance: start
with 100 infectious units divided among members of a single litter of
10 newborn mice and in two weeks they will give enough virus to
infect 10,000,000 or more mice—or quite a few humans. Certainly
there's some hazard, particularly when the mouse brains are put in
the blender for processing or the sprayer they would use to generate the aerosol, but on the whole, it's pretty safe—a lot safer than
being the driver of a truck full of explosives in front of the Marine
barracks in Beirut.

Yes, there's a vaccine for tick-borne encephalitis. But how many
people have had it? Other potential agents I would lump together
with this are anthrax, some of the hemorrhagic fevers, VEE, plague,

tularemia, and brucellosis. They all grow readily to high titer, most are pretty stable in aerosol, and the methods of cultivation are easily accessible in the open literature. Many agents would tie the nation's agriculture into knots.

Smallpox is a terrorist's dream. The virus no longer exists in nature, but monkey pox, another pox virus that infects humans, has been acting up in central Africa. Certainly smallpox exists in freezers in Atlanta and Novosibirsk; other countries claim to have destroyed all their stocks, but there is no mechanism for inspection or verification to be sure. Most of the population these days is susceptible, it grows easily in mice, eggs, or cell culture, it's very easy to disseminate, will spread like wildfire to others, and has a high mortality rate. Our stockpiles of vaccine are low and we don't know how long the vaccine remains viable.

While botulism has a small area of effectiveness compared with infectious agents, it's deadly if used to poison a small water supply or sent through the air-conditioning system of a building.

Ricin, a poisonous protein, is easily purified from castor beans. Staphylococcal enterotoxin B won't kill people, but it'll make them feel like they want to die. It will knock a person flat, so nauseated he can't move his head. It was developed by the U.S. Army as an incapacitating agent. It's very stable and highly toxic by aerosol. You can also get it in bad potato salad and, believe me, you'll never forget it if you do.

Chemical terrorism has already begun, and biological terrorism is unlikely to be far behind. When members of the Aum Shinrikyo cult set off a device releasing the nerve gas sarin in a crowded train station in Tokyo in 1995, their actions struck fear in urban areas worldwide. If it could happen in Japan, an affluent society well known for its low crime rate and social cohesion, it could happen anywhere.

More and more often, it seems, when something like this happens—or when a suspect is arrested following the blast in Oklahoma City—intelligence sources are quoted saying something like "They weren't even on our radar screen." What this indicates to me is the need to focus on the threats that are already present and to throw the net more widely to pick up new problems creeping out of the woodwork. It's just like the need for emerging virus surveillance. Many in international circles had heard of the Japanese cult, which also had ties to Russia, but the weaponization of chemical agents caught most unawares. I fear that this represents just a chilling example of what we can expect in the future. Nation-state terrorism will be augmented by a complex mosaic of paramilitary crusaders, religious true believers, and aggrieved splinter groups, many of whom will tap new and innovative talent to create their weapons, sometimes with assistance from active or defunct national BW programs.

Tokyo commuters were actually fortunate in several respects. First, the cult's dissemination devices fizzled. There could easily have been deaths well into the thousands. Second, as bad as the effects were from the nerve gas attack, the theoretical and practical problems associated with a biological agent would have been much greater. And Aum Shinrikyo cult members actually went to Zaire looking for Ebola virus.

There are rapid field detectors for sarin which would have allowed police, medical, or fire personnel on the scene to test for the presence of the agent and its dissipation. Tom Ksiazek has shown that relatively simple immunological tests in the laboratory can readily detect most of the hemorrhagic fever viruses as well as anthrax toxin in human blood, but there has been little significant further developmental work to produce field-expedient tests applicable to biological weapons—tests which would function much as the immunological pregnancy tests which are so sensitive and reliable. Since the infectious agents multiply in the host body, it would only be necessary to infect one target, who would then walk away from the scene of the "attack" unaware of infection, incubating the disease. No alarm would be raised and there would be no warning.

The immediate effects of a sarin nerve gas attack are dramatic and horrible, but the long-range effects of a silent attack with biological agents could be far more crippling. If thousands of daily commuters in Tokyo were to leave a train station not knowing they were incubating a lethal infection (and, in the case of some biological agents, capable of infecting their loved ones, co-workers, and more), imagine how many more potential targets they could reach than the dozens killed by sarin. The effects could be far-reaching, depending on the agent used and any long-term permanent damage to survivors.

In the United States, we are no better prepared to deal with an actual attack than the Japanese, as was proved in a drill carried out in New York. Confusion ensued and overzealous firemen entered the drill area without protective gear—a mistake that would have been fatal in the situation in Tokyo. This is particularly ironic given that the drill was conducted in the same city where the army once ran its own test of releasing surrogate biological weapons in a subway system!

It seems almost naive that some American legislators fixate on large-scale defense concerns, including billions of dollars in sophisticated missile systems like the Patriot to protect us from the likes of a Saddam Hussein. We've already seen how one rented truck in Oklahoma City filled mostly with manure can bring the country to its knees. In fact, a threat greater than both of those could fit in the back of a pickup truck.

Now for the real scary stuff:

In the words of Nobel Prize-winning geneticist Joshua Lederberg: "Visualize the World Trade Center or an Oklahoma City-style attack complicated by the inclusion of a kilogram of anthrax spores as a kind of microbiological shrapnel along with the explosives. And [imagine] its implications for salvage and rescue, public health, panic. If I just mention the word Ebola, you have some idea of what I am talking about."

The U.S. Office of Technology Assessment has calculated that a light plane flying over Washington, D.C., carrying a hundred kilograms of anthrax spores and equipped with a standard crop sprayer could deliver a fatal dose to around three million people. Enough anthrax spores to kill five or six million people could be loaded into a taxi and pumped out its tailpipe as it cruises the streets of Manhattan.

Unlike nuclear weapons or even many incendiary-type homemade bombs, biological weapons can be grown easily and quickly with little more than a home-brewing kit. For small nation-states sponsoring weapons development, why wait the decades and spend the bucks it can take to develop a nuclear warhead—painstakingly assembling ingredients while the entire world keeps track of your progress—when it is so much easier and faster to obtain the materials for a biological weapon and they can get results with about five average biologists? We dodged a bullet in March 1995 when someone at the American Type Culture Collection, a clearinghouse for bacterial cultures, became suspicious about a request for shipment of *Yersinia pestis* samples—the bacterium that causes bubonic plague. Ironically, the microbiologist and member of the Aryan Nation white supremacist group who requested the cultures (which were to be sent by mail) got in trouble because he forged a research firm's letterhead. His attempts to obtain plague, however, were completely legal.

Do-it-yourself biological weapon making is not without risk for the developer, obviously, but most fanatical terrorists are willing to take tremendous personal chances, particularly when the potential for damage—physical and psychological—is high. As we saw in the

Tokyo subway attack, these types of homemade weapons are unpredictable. This very unpredictability could work in the terrorist's favor.

The solutions to the problem of biological terrorism are much the same as for combating threats of disease coming from nature. Effective surveillance must be a high priority. In Japan, investigators were horrified to learn that several members of the Aum cult belonged to one of the country's elite military units. Similarly, U.S. military officials were shocked to find several extremist militia members in the ranks at Fort Bragg. Somehow, without creating a "big brother" atmosphere, we must find a way to identify the sources of these threats and get them "on our radar."

Surveillance must be global, just as it should be in the world of wild-caught infectious disease. Terrorists exist in every country. The Aum Shinrikyo was an international cult, with members worldwide. The tools and practices of terror are not restricted to any particular culture. As use of the Internet continues to proliferate and the world effectively grows smaller, we can only expect that information about and access to such weaponry will be swapped more easily.

The best protection against this type of threat may be aggressive education of physicians about the earliest symptoms of common agents, along with the information and means to diagnose and treat injuries from both chemical and biological weapons. We can't afford the delay in recognition of an index case if we hope to be on top of what could be mass casualties. Once an agent is identified, the medical experts cannot be in the position of searching through textbooks to find the best therapeutic approach. Even if we are forewarned, there are often limited preventive or therapeutic choices for the diseases we are concerned with, and the deployment of those we have has not been thought through. The military is tightly focused on anthrax and botulism, but other programs are suffering. These include the hemorrhagic fevers, which are both biological warfare and natural threats. As with natural epidemics, we have to be able to recognize what we're dealing with and know ahead of time how to react.

We had assembled a group of experts, including Bill Patrick, to teach a BW defense course at USAMRIID, and now, after the Gulf War, it has become SRO and seems to be having a broader impact among decision makers. In the civilian community, where I now live and work, we are, unfortunately, woefully behind the curve. Before a 1995 Senate hearing on the matter, Office of Emergency Preparedness official Frank Young warned, "There is no coordinated public health infrastructure to deal with the medical consequences of biological terrorism."

In the worlds of virology and epidemiology, scientists often speak in military terms of the need to be prepared. We can't always be "fighting the last war." Whether the issue is an outbreak of an unknown disease within our nation's borders, a chemical or biological attack in one of our megacities, terrorist contamination of our food or water supply, we have to be prepared to deal with it at every level: intelligence and surveillance must be carried out in an effort at prevention; the first-response teams of police, fire, and rescue must be trained in how to respond to various such emergencies; hospitals must likewise have the training and procedures to mount an effective response; and physicians and scientists must be trained to recognize and take care of the victims. We have no other options but to prepare.

It wasn't very long after the troops came home from Desert Storm before we at USAMRIID were caught up in the same budgetary cuts as everyone else. I probably could have stayed, but I didn't want to stay without a significant goal. The Disease Assessment Division had essentially been created back in 1984 to give me a place to pursue the studies the U.S. Army and I both thought were important. If that wasn't going to be true any longer, it looked as if my work for the army was done. Gerry Eddy was gone. Dick Barquist and Dave Huxsoll were gone. It was time for me to go too.

But infectious disease is a small world. It so happened that CDC was looking for a chief of the Special Pathogens Branch, Division of Viral and Rickettsial Diseases of the National Center for Infectious Diseases. The position had originated with my mentor, Karl Johnson. When he left, he'd been succeeded by Joe McCormick, and subsequently Ken Herrmann became acting director.

Fred Murphy, the head of CID, called to ask me if I'd be interested in applying for the job. Susan and I really liked our life in Frederick, and I wasn't anxious to leave the army, but this seemed like the natural position for me. I would be doing much of the same sort of thing for the civilian sector that I had been doing for the military in USAMRIID. I felt I'd made my case in the army, and I wanted to see if I could pursue the same goals in Atlanta. Just as Gerry Eddy had lured me to USAMRIID years before with the promise of all that fetal calf serum, this time the lure was the only other Level 4 lab in the country, recently completed in a gleaming new building. Okay, they'd scrimped a little on office space and mine would be in one of the many basement levels without any windows, but creature comforts had never been one of my big motivations.

So we moved to Atlanta, where Susan became the Coordinator of Language and Literature for the Emory University library system, working not far from my own office on Clifton Road. Shortly after we settled in Atlanta, we sold the old Toyota in which I'd transported the hot frozen monkey corpses back to Detrick. We happened to spot the car a year later in the parking lot of a local shopping center and wondered what the new owner would think if he knew what a role that vehicle had played in an episode that was almost our first home-grown biological natural disaster and had become the subject of a number one best-selling book, *The Hot Zone*, by Richard Preston.

Before I left USAMRIID, I was given an honor that meant a tremendous amount to me and showed me that despite the fact that I sometimes felt otherwise, I wasn't just a candle in the wind. USAMRIID had established an award in my name:

THE C. J. PETERS AWARD
for Excellence in Fulfillment of the Mission
of the United States Army Medical
Research Institute of Infectious Diseases

Underneath the USAMRIID insignia with its motto, "Research for the Soldier," the plaque reads:

In the inception and granting of this award, we honor Colonel C. J. Peters, Medical Corps, a brilliant and innovative scientist, scholar and physician, and an example of unfailing dedication to the protection of the health and well-being of our Armed Forces. The recipient of this award follows in the tradition of Colonel Peters' persistence and creativity in the search for prevention and therapy of militarily relevant diseases, and reflects his enthusiasm for surmounting the obstacles inherent in the struggle to advance medical knowledge. Most important, the recipient of this award is a mentor, inspiring younger researchers to participate and succeed in the challenge and to share that profound reverence for life that is illustrated by the career of Colonel Peters, whose work is a lesson not only in science, but in humanity.

I was moved and humbled. If half of what it said about me was true, then maybe I had accomplished some of what I set out to do.

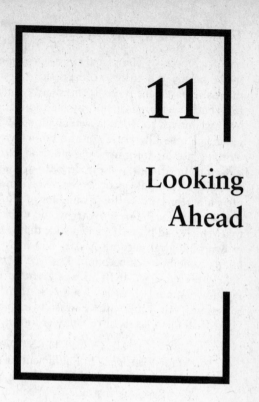

11

Looking Ahead

THE YEAR 2012—*The Roots of an Outbreak:*
The government health center in Bangkok is, like the city itself, a hectic, busy place, teeming with people. Mondays are always the worst at the clinic, when people with problems from the weekend overflow into a cramped waiting room too small to handle a normal day's schedule. Fortunately for both doctor and patient, most people come in with predictable complaints: mothers bring in children with earaches and sore throats; the elderly come in with pains and coughs; occasionally a cut or broken bone needs attention; prostitutes and others come for treatment of sexually transmitted diseases, including AIDS, or confirmation of an unwanted pregnancy. Finally, there are people of all ages suffering symptoms related to poor hygiene and inadequate nutrition in general, sometimes complicated by overconsumption of alcohol and drug use.

It is fortunate that the diseases and symptoms are predictable because there is no time to take a complete medical history of everyone who walks in the door. The best the staff of the clinic can do is spend five minutes or so on a quick once-over, sometimes administering a shot of antibiotic or writing a prescription for oral drugs, before moving on to the next case.

One Monday is even busier than usual. In midmorning Dr. M. sees a young woman and her four-year-old son. The boy, obviously feverish, was lethargic all weekend and wouldn't eat. Confirmable symptoms include the fever, malaise, and myalgia, and a slight cough. The doctor recommends over-the-counter drugs for the child's fever and reassures her that children often develop colds with fever. He tells the mother to bring him back if he isn't better in a few days.

In this overcrowded, understaffed, and undersupplied clinic, there's no such thing as a routine test. If the mother belonged to a different social stratum and her child was seen at the private office the clinic doctor worked at in the afternoons, perhaps he'd take a throat culture or prescribe an antibiotic—not so much because he suspected the child had a particular infection but to make the mother feel that something was being done. At the clinic, though, people were usually so grateful for help and attention that assuaging the client's feelings was not an issue. Tests were not available and drugs were prescribed based on what was on the shelf. An antibiotic still within patent never showed up there. Most of what the doctor does is based on experience and calculated guesswork.

The next day he sees the same woman, now sick herself, and coming alone. Her little boy is worse. Aspirin doesn't seem to help his fever and his cough is more persistent, and he is bringing up a lot of sputum, often tinged with blood. He is having difficulty breathing. She, too, now has a cough. The doctor tells the woman to bring the child in the next day and gives her a shot of an inexpensive antibiotic along with a prescription for a cough suppressant.

The mother and child do not show up the next day, though the doctor is so busy he doesn't remember them until he is home that night, relaxing with his own son. He worries they may have rapidly progressive tuberculosis. He hopes this is not the case, though, since he doesn't have access to a combination of drugs that will kill the resistant strains prevalent in the city and now common in many places throughout the world. He makes a mental note to have one of his assistants follow up in the morning.

The next afternoon the assistant from the clinic goes out to check on the woman and child. The address she provided is on the fourth floor of a large, run-down apartment building. After waiting five minutes for the elevator, the assistant takes the stairs, despite the stench of urine. A knock at the apartment door is answered by the *woman's sister who, looking bereaved and terrified, says both her sister and her nephew are dead.*

Now the whole family is worried about her brother-in-law, who is sick with fever, bloody cough, and prostration. Of all the members of the family, his health is the most crucial, as all eleven family members living in the two-room apartment depend on his income. If he does not get well soon, they won't be able to buy food. The only saving grace is that his fever seems to be distracting him from the loss of his family.

When Dr. M. hears this news on Thursday morning, he is stunned. How could the child die? How had the mother's illness progressed so quickly? One thing was clear: the woman's husband has to come in for diagnosis and treatment immediately. Further, with three members of the family infected so rapidly, the disease was likely spread by person-to-person contact and/or inhaling whatever the sick were coughing up. With the crowded and unsanitary conditions in the apartment building described by the doctor's assistant, it would be very easy for the disease to spread to others in the family and even outside the apartment.

The doctor sends the clinic worker back to the apartment on Friday, instructing him not to return without the patient, but learns the patient is too sick to travel.

By Monday morning the man has already died and been cremated. Family members usually do not allow autopsies of their loved ones, and even if they did, the clinic had neither the proper facilities nor the resources to perform one. So the doctor is left with three dead patients, with no firm cause of illness or death and no samples to analyze.

Clearly, this was worse than any variation of bronchitis or TB he's seen. At home that night, the doctor turns to the Internet, taking the clinical description of his patients on-line to see if any of his colleagues can provide any insight. One, a hematologist, has seen two patients recently with similar symptoms. Like the clinic patients, both came from crowded, poorly ventilated living quarters in a slum area. The hematologist performed complete blood counts and noted the hematocrits were higher on these patients than he expected, but otherwise saw nothing significant in the blood.

As he kisses his son good night that evening, Dr. M. can't help but think of the young boy who died, wondering if he could have done anything different in the five minutes he had with him—if he should have seen something to indicate just how sick he was. The clinic doctor forwards the information from the hematologist with the description of his patients in a letter to the Ministry of Health by E-mail.

The Ministry of Health has no resources to look into small case clusters. After all, sickness is a fact of life in the poor sections of the world's megacities. This week, however, a new class of epidemiologists are coming up through FETP, the Foreign Epidemiology Training Program, originally established by CDC and now maintained by the Thai government. Looking for projects for his trainees, their mentor reads the clinic doctor's notice and assigns one of his pupils to investigate the cases.

The trainee takes three days to track down everyone who had contact with the sick but finds thirty-five people in six units of three different apartment buildings who meet the case definition. With her supervisor's help, she designs a case control study that indicates that close contact is the only detectable risk factor. Using a more detailed follow-up questionnaire, she determines that physical contact or common meals are not required; proximity to the patients seems to be enough to spread the disease. All the patients share conditions of extreme crowding and poor diet and hygiene. It takes the trainee two weeks to analyze the data from the study. By then, more people have taken ill.

The Thai Ministry of Health recognizes they have an epidemic in progress in one of the region's largest cities—a potential disaster. But they have no resources to do an in-depth epidemiological study, and no lab to look for a specific infecting agent. Public health officials turn to the U.S. Army laboratory in Bangkok for assistance. Together, they still cannot identify a causative pathogen using tests designed by the U.S. military and the World Health Organization for agents causing respiratory disease in the tropics.

There are no informative notices on WHO's worldwide net, and the Thai government is reluctant to use the web for anything other than research. They fear that posting information on the outbreak on an open site will cause panic; the news agencies monitor the net for just this kind of story. WHO experts agree that a media circus would be counterproductive to an investigation but have limited resources to add to the investigation and control efforts.

The Centers for Disease Control is also contacted. Case count and symptomatology are compelling enough for them to send a team to investigate. With limited resources itself, CDC can spare only a handful of people, chosen for their expertise and experience. The team from CDC works with the local health agencies to canvass the neighborhood where the first reported cases lived. Alarmingly, they find that more than 250 people have fallen ill. The epidemic is so severe that the Thai government, WHO, and CDC discuss the possibility of quarantine; they decide there isn't enough data as yet and that it would be unlikely to be effective in any case.

In the field, scientists observe clustering of disease within rooms and learn from locals that the mortality rate is more than 50 percent. As the size and scope of the problem grows, more experts are dispatched from CDC, including several with respiratory pathogens expertise, as well as a pathologist. Samples from the sick and dead are sent back to Atlanta for culture and molecular biological analysis.

Initial testing for influenza and known respiratory pathogens is negative. However, histopathological examination of samples from four fatal cases shows that three of the victims have cell changes suggestive of a virus that causes cell fusion and death; it looks a lot like something the CDC had seen back in 1994, from cases of a mystery paramyxovirus from Australia. That is worrisome; because the 1994 outbreak was restricted to just a few cases and hadn't surfaced since, little was known about diagnosing or treating the virus. In-depth study of that virus was also slowed by powerful economic interests in Australia, fearing their tourism and export industries would suffer.

FETP personnel return to the area of infection to draw blood samples from people in contact with the sick. In one of the apartment buildings hardest hit, they notice increasing transmission in one particular room. Over the past three weeks, the rate of disease seems unusually high, and these later cases seem to have had much more casual contact or exposure than the earlier ones. The FETP workers hypothesize that the character of disease transmission has changed, a hypothesis which is supported by the evidence. The disease is now clearly spreading by aerosol, and on average, each case generates several secondary cases.

Three months after the original doctor saw the first sick child, the disease has spread throughout the poorer sections of the city, with pockets of infection exploding, particularly in massage parlors and among the children of mothers who work in them. These are spots frequented by tourists, and soon infected businessmen bring the disease home with them. Small clusters of cases break out in Germany, Japan, and Scandinavia, although initially these go unnoticed.

Looking Ahead 307

The Thai government has still not agreed to post notice of the outbreak internationally on the Internet, fearing political and economic repercussions. It is not until the epidemic has grown so big that the international news media pick up on it that cases are linked worldwide.

This imaginary scenario is not meant to be prescient of a particular epidemic in the future, but illustrative of conditions and forces that already exist and will affect the next disease outbreak. For while I don't pretend to be able to predict what the next great disease will be, where it will strike, and how many people will die, or how we will combat it, I believe it is possible to identify some of the most dangerous pieces of the puzzle beforehand.

Deadly infectious diseases have been a hot topic both in the world of the public health scientist and in popular culture for some time now. For the virologist or physician specializing in infectious disease, the interest has been motivated by necessity. After decades of gains made as humans have invaded every corner of planet Earth, it seems the microbes are making a comeback. In the past twenty years we've seen thirty or more "new" diseases surface around the world. In that same time period, dozens of the old standard diseases have reemerged: diseases like tuberculosis, yellow fever, malaria, and cholera.

After years of being off the public radar screen in the face of killers like heart disease and cancer, infectious diseases have once again dented the public consciousness. Whether the reports convey dramatic, exotic, blood-dripping death from Ebola in Africa, the overwhelming tragedy of AIDS, or concern over safe food, infectious disease—the leading cause of death worldwide—is no longer buried in the back pages of the newspaper.

The outbreak I have sketched is based on a real paramyxovirus only recently identified and its reservoir even more recently traced to a species of fruit-eating bat in Australia. Little is known about its natural history, and the only concerted research being carried out on the virus is in a single laboratory in Australia. It is likely that related viruses exist elsewhere, both in rural and urban areas. The scenario is similar to the one that we believe resulted in the emergence of measles virus (also a paramyxovirus) as a human pathogen about 5,000 years ago, only somewhat compressed. The facilitating event for measles to make the jump from an animal reservoir to humans was the development of irrigated agriculture, which in turn allowed hun-

dreds of thousands of humans to congregate in cities; in our hypo-
thetical case, the megacity was imagined to be the trigger.

I have not taken the worst-case scenario. I described a situation
where some medical care existed for even the poorest and where
communications and movement were still possible within the imagined megacity. Globally, there are many areas today where there is
essentially no recourse to medical care and where epidemiologists
would not dare to enter. Already, in Manila, businessmen rely on
helicopters to move between the tops of tall office buildings to escape
the traffic gridlock and danger of the streets. There is no sign that
these trends will change, and indeed, with exponentially climbing
population figures, many poor countries will double in population
within fifteen to twenty years and neither growth in earnings nor in
food production will keep up.

Nor have I taken the gloomiest case in biomedical terms. I have
supposed a level of surveillance, discovery of disease, and response
that is more efficient and rapid than exists in Third World countries
today and optimistically assumes that current plans in the international community will be implemented. But much more needs to be
done to prepare.

Let me try to give you an idea of how I see the situation facing us.
First, you have to readjust your view of the world. If you are like
most people, you see yourself and those around you as the main
players in the world you live in, complemented by a few horses, dogs,
and cats. For the most part, the animals and plants around you
"belong" where they are; they will be there in the future except for an
occasional spotted owl or other unlucky creature that somehow
doesn't do so well for itself and is on the brink of oblivion, maintained only by the expensive but resourceful efforts of conservation-
ists.

Such a world view is entirely mistaken. First of all, humans by no
means occupy center stage in the biosphere. There are only 4,000
mammal species in the world. By far the largest number of living
creatures are the insects, numbering 750,000 species. Even beyond
that are vast countless numbers of microorganisms. Bacteria, for
example, inhabit the most common as well as the most unlikely
places in the world, including molten rock under the earth's surface,
hot geysers, ocean depths both hot and cold, strong acid solutions,
and so on. A teaspoon of soil may harbor 100,000,000 bacteria, and
a square inch of skin more than 100,000 of them, all serving a useful

purpose in the earth's or the body's economy. We can't even begin to count the number of viruses around. They parasitize bacteria, plants, and all sorts of animals; they even grow in the common protozoal parasites in our gut, such as the amoeba. We have found more than 600 different viruses circulating in nature in arthropods or rodents with potential pathogenicity for humans, and this is only relying on the old classical tool of virus isolation applied in a few labs around the world that had a particular interest in looking for them.

I suspect most readers also need to rethink their ideas about the stability of the world around them. The biggest species changes in recent U.S. history were probably seen with the Columbian interchange—cattle, horses, and pigs were brought to the Americas after 1492, as well as the wheat for making bread displayed on supermarket aisles. However, these visible and useful imports are minor compared to the dandelions, clover, and other European weeds and grasses which we don't even recognize as foreign today but which have transformed our landscape. These intruders continue at an ever-increasing rate as commerce and travel accelerate—kudzu, killer bees, fire ants, exotic mosquito vector species such as *Aedes albopictus,* and so on.

Many of the disease agents we now accept as part of everyday life were brought from the Old World, decimating Indian populations, facilitating the conquest of the original inhabitants of the Americas, and making it possible for the Europeans to build settlements. We have mentioned the movement of plague and yellow fever as examples, but measles, diptheria, smallpox, and many other diseases also made the crossing of the Atlantic.

We have also seen how some of these movements have led to unexpected problems. The importation of horses and burros into South and Central America provided a dangerous amplifier of Venezuelan equine encephalitis virus, which we think was probably circulating among rodents and birds until these big, susceptible hosts gave it a chance to evolve into the epidemic scourge it has become in this century. Cattle and sheep brought to Africa a few thousand years ago, reinforced by the more recent importation of large numbers of highly susceptible European breeds, have served a similar role in the amplification of Rift Valley fever virus transmission. It is interesting that these viruses both seem to still be dependent on these newly (in the evolutionary time scale) introduced amplifiers to cause epizootics. Nonetheless, they are capable of reaching high levels in human blood and must be watched carefully to see if humans become a substitute for the domestic animals in amplifying epidemics.

While these plant and animal species are moving about the globe, the viruses are rapidly evolving "new" species. VEE, in particular, seems to be unstable. If Scott Weaver and Becky Rico-Hesse, molecular biologists at the University of Texas and Yale, are right (and I think they are), this virus has spread over South and Central America in the last 1,500 years and has evolved into different varieties of virus in different environments. The VEE virus that seems to be forcing itself into the horse population in Colombia and Venezuela is the most notable, but we have a VEE strain in the Everglades that probably arrived only 150 years ago and could emerge from its natural environment by attacking a new host.

Species are disappearing as well. Maps of the earth's surface made from satellites show that humans now have radically changed half of the lands available. As we destroy long-established habitats, we are bringing about the loss of our comrade animals and plants on the earth in a way that threatens to equal the level of extinction seen 65 million years ago, when the dinosaurs died out. But destruction of one creature's habitat creates a place for others, leading to movements of native or introduced plants and animals into new areas, and juxtapositions of species never before encountered. The loss of the spotted owl is in itself not particularly important, but its possible demise is a signal that an entire ecosystem is in critical condition.

These changes are all leading to a more complex and rapidly shifting mosaic of natural ecologic niches, combined with the change of human habits, which are also just a part of the viruses' ecology.

But if the fluidity of the natural world to create new receptive "homes" for viruses presents a formidable challenge for prediction and control of diseases, the viruses are also capable of change. In one of the more predictable elements of the scenario opening this chapter, the virus undergoes one or more genetic mutations. In fact, virtually the only given in the world of RNA viruses is change. As viruses reproduce and random mutations take place, most have little or no effect on the virus in the long term: alterations that make the virus less viable disappear, whereas changes that are beneficial to transmission are selected for naturally as the "fitter" virus replicates itself.

The question then becomes how different from the parent virus can these mutants be? For example, we know there has been a polio virus at least since the time of the pharaohs because we have bones that show impressions of the typical skeletal deformities that follow the nerve damage and consequent muscle atrophy resulting from polio virus infection. There must be something that keeps polio virus as polio virus, hence there must be limits to the changes that are compatible with survival of these viruses.

Generally, new viruses retain some of the properties of the viruses that they came from; and within the same family of viruses, even after thousands of years of evolution, the individual viruses retain certain themes. All known arenaviruses, for example, are transmitted between rodents and often have some specialized mechanism of transmission to the next generation of their particular rodent species.

One important change seen in evolving viruses is in antigenicity. Only influenza viruses make a habit of this, changing their antigenic composition from year to year under the pressure of the immune system of their human reservoirs. As the population of humans in a given area develops immunity to a particular flu virus, the mutations that will be successful are those that change the antigenicity enough so that it will be unrecognizable to the hosts' immune systems.

There can also be bigger antigenic changes that can allow influenza viruses to circumvent vaccines and immunity from previous infection in the host. To make these dramatic changes, influenza virus obtains an entire new gene that the human population has not experienced. For the last fifty years these new genes are thought to have come from waterfowl in Asia and resulted in a worldwide outbreak of influenza.

Just a small change in the genetic makeup of a virus can result in tremendous differences in the behavior and character of the virus—although this is not the most common cause of an emerging virus disease. People with dogs are advised to inoculate puppies against canine parvovirus, for example, which causes a disease that can be fatal to dogs but does not infect humans. Canine parvo actually comes from a mutation of a cat parvovirus, which probably "leaked" from cats or wild animals. The genetic code of the original virus changed just slightly, like a change in eye color from blue to green, but with tremendously deleterious results for the dogs it was suddenly capable of infecting. Once the virus got in dogs, the changed virus was selected for and propagated from dog to dog. The disease causes severe diarrhea, with infected dogs literally wasting away. This mutation has led to the death of thousands of dogs in what we call a pandemic—an epidemic that travels across the world.

Some may argue that cats and dogs are so different from humans that it is not worth citing this example. But I think the lessons are important. From the virus's standpoint, any organism is as good as the next: cat, dog, human, mosquito, or flea. The fact that we have opposable thumbs, know how to use vacuum cleaners, and listen to opera simply doesn't matter to the viruses.

I've talked mainly about viruses that are transmitted to humans by a vector, an animal transmitter such as direct contamination of the environment by an infected rodent or the bite of an arthropod such as a mosquito. The vector, or intermediary, is beyond the control of the virus and can be critical to the "success" of the virus' transmission.

In the case of a mosquito-borne virus, for example, the "fit" of the virus to the mosquito is critical and many subtle environmental factors enter into the ability of the mosquito to reproduce and to reach its own hosts.

Aedes aegypti is a good example. This mosquito was brought from Africa to the New World and was responsible for the establishment of yellow fever and dengue as plagues on our lands. It was possible to control it after 1900, which, in turn, controlled yellow fever. But a variety of scientific and social factors have seen a recrudescence of the pest on the Gulf Coast and in Mexico, the Caribbean, and South America, with epidemic dengue and dengue hemorrhagic fever as a consequence—and the real risk of a return of urban yellow fever, an unhappy circumstance I'd predict in the next decade in the Americas.

One of the big ecological/evolutionary crapshoots that we are witnessing today is the outcome of introduction of the Asian tiger mosquito, *Aedes albopictus,* into the Americas. This mosquito, which is an efficient vector of dengue, was brought to the United States by shipments of scrap automobile tires, and like many immigrants since 1492 it found a very hospitable landscape. It resembles *Aedes aegypti* somewhat but is more flexible in breeding sites and climate preferences. There was no concerted effort to eradicate it after its arrival in Houston in 1985, and it is now genetically adapted to its new home, is found from Chicago to Florida, and has become a major man-biting pest mosquito along the Gulf of Mexico.

The mosquito is an efficient transmitter of many arboviruses from North America and elsewhere around the world when it's tested in the laboratory. Will it make Houston a receptive area for dengue, or will it be able to increase LaCrosse encephalitis in the middle western states? We are simply waiting to see if the ecology of the viruses and the mosquitos bring them together over the next few years. The most unpleasant threat is a disease called eastern equine encephalitis, or

EEE. Like many vector-borne viruses it circulates annually in a cycle that only rarely involves humans; the mosquitoes live in swamps where there are relatively few people and the mosquitoes themselves are not intensely fond of humans for their blood meals. In 1991 EEE infected *Aedes albopictus* mosquitoes were found in Florida. The Asian tiger mosquito was so named because of the tiny stripes on its body, but it is also a tiger when it comes to biting humans. EEE is an extremely virulent disease. The mortality is 50 to 75 percent and the survivors are often left with physical or mental impairment.

So as you can see, we in the United States are performing a big, complicated, irreversible experiment with this mosquito. I'm sorry, but nobody can tell you the isolation of EEE from the Asian tiger mosquito was just an incidental finding that has no sinister implications for the future any more than I can tell you that there will be future problems.

As an example of how far away we are from accurate predictions of this kind, I recall a presentation on LaCrosse virus and its vector mosquito. After several equations and careful considerations of the ranges of variables, the scientist declared, "I conclude that LaCrosse virus is impossible." He was partially joking, but I was reminded of the story of the aeronautical engineer who made a series of calculations and concluded that it is aerodynamically impossible for the bumblebee to fly.

Nor can we really determine which virus will "fly" in terms of new diseases. The virus in my imaginary scenario is modeled on an obscure member of the paramyxovirus family. Although we don't know much about the virus itself, it has distant relatives in the family that have shown a propensity to jump species, including canine distemper virus that has recently been noted to kill lions in East Africa, as well as the precursor virus of human measles. The virus that caused the outbreak in Australia jumped species from a fruit-eating bat and was capable of acutely infecting and killing horses, humans, and cats. The three reported human cases were severe respiratory diseases that occurred in people who had close contact with infected horses. Paramyxoviruses are also quite transmissible from host to host (measles is highly aerosol-transmissible).

The main difference between the virus of fact and that of fiction is that in the real outbreak there was no person-to-person transmission. My guess is that if you tried the experiment often enough in the crowded, unhygienic setting of a megacity, the virus would eventually evolve into a form that permitted person-to-person transmission.

Why can't we predict more about the outcomes of cross-species transfers, mutations, vector introductions, and other such changes in our biological world? We can do a lot better than we could in the past, but no increase in the number of molecular geneticists, epidemiologists, disease ecologists, and others will ever completely fix the problem because of the concept of chaotic systems, which applies in spades to viruses. Modern chaos theory actually began with the study of weather when MIT professor of meteorology Edward Lorenz realized from his atmospheric models that tiny changes in initial inputs had huge effects "down-line," doubling and doubling again every so many units of time and distance. He summarized the concept with his now famous question: "Does the flap of a butterfly's wings in Brazil set off a tornado in Texas?" What Dr. Lorenz ended up concluding was that, contrary to earlier optimistic projections, we would never be able to predict weather with high accuracy on anything other than a very short-term scale because of the chaos inherent in the system.

I'm afraid chaos theory has the same profound implications for medical science. There is one valid, simplified equation to describe what we can expect from viruses in the future: Variable Viruses + Changing Ecology + Increasing Movement/Transportation = More Infectious Diseases Developing at an Increased Rate.

The existence of each element of this equation is a given in modern times, and there are endless variations on possible parameters for each. RNA viruses in particular, as we've noted, make "mistakes," or mutations, all the time. Within the same family, however, viruses tend to retain some similar properties as they evolve. This is why, when people speculate about HIV mutating to become aerosol-infectious, while you "never say never" in virology, it's very unlikely. None of the retroviruses (the family to which HIV belongs) has shown that property either in nature or in the laboratory. This kind of argument by analogy can be useful in guiding our explorations of the unknown, but you can't "take it to the bank."

With each new virus I've studied, I've developed even more respect for how well they do their thing, despite any "sloppiness" they may show in genome replication. In some ways, they are the ultimate parasite, ideally adapted to whichever reservoir, vector, or host they use to forward their cause. Take canine rabies, for example. The virus gets to the salivary glands of the infected animal, waiting for a ticket out to the next host. At the same time, in the brain the disease is making the infected animal mad so that it will bite anything near it in a blind fury. In other words, the virus causes a disease with symptoms that create the ideal delivery system for propagation of more virus. This kind of ecologic adaptation is at least as subtle and complex as the matching of a protein on the surface of a virus to the virus receptor on the cell it is infecting, and demands a completely different kind of analysis.

When you consider how much thought, planning, experimentation, sacrifice, blood, sweat, and tears go into our attempts to understand and outwit them, the viruses' unthinking, tenacious selection and reproductive efficiency is both impressive and humbling. Microbes have continued to thrive (and gain ground) despite our best efforts. A half century after his discovery of how bacterial mutations evade antibiotics, Nobel laureate Joshua Lederberg predicted: "I see this as a race that will be settled in fifty years . . . It'll either be our technology or it will be the evolution of the germ world that will have established its ascendance."

It's difficult to get people to see such evolution as it really is— absolutely morally neutral. Some of it will be positive for any given species or life form and some of it will be negative. I don't think we can afford to be arrogant as a species. We're subject to the same rules as everyone else. We're not going to be exempt from the process of evolution. Looking at the Mayan and other past civilizations which are represented by the ruins I like to visit, it is clear to me that we can't be complacent. History and evolution are actually against us.

Diseases in general, and viruses in particular, have always had an enormous impact on natural populations, a fact most clearly exemplified by the European-American interchange at the time of Columbus. It wasn't just all those scary horses, it wasn't just all those deadly firearms that allowed a handful of mean, tough Spaniards to knock off an entire continent. It was the diseases the Spanish brought with them.

Viruses are mindless in their random selection. Another way to say this is that they're just about as shortsighted as humans are; they don't have a long-term strategy, just a lot of short-term strategies. They cross species, mutate, run virulent through a tribe or group, and then may lose themselves by killing off or immunizing all the available hosts. But occasionally, when conditions are right, they establish themselves and humans have a new challenge.

We tend to have an anthropocentric vision of the way the planet works. Even when people worry about the way we are altering the environment, they tend to think of it in terms of how these changes will affect us without recognizing the true scope of the perturbations.

In part, this is because, true to the chaos theory, we cannot predict all the actions involved in our ecological interactions. When you add the moving target of ever-evolving, ever-mutating viruses, it is impossible to gauge how the actions of man will impact on emerging infectious diseases. How many alterations in nature occurred as a result of the construction of the Aswan Dam? Did, for example, its destruction of the floodwater mosquitoes' habitat help terminate outbreaks of Rift Valley fever in Egypt? On the other hand, the alteration of the yearly inundation patterns of the surrounding fields profoundly changed the Nile's ecosystem in countless other ways, including radical shifts in human parasitic infections but improvements in nutrition.

You don't have to be an environmental activist or "tree hugger" to know that we are massively changing the natural state of the planet. It's hard to find sizable primary forests any longer unless you go to the Amazon basin or central Africa, and even there the rain forest is being cut down every day as we relentlessly encroach on unique pieces of nature's real estate.

Altering natural habitats can create new ecological systems, allowing different species of flora and fauna to move in and take over, usually simplifying the local ecology. When you clear a forest area for farming or cattle, the rodents indigenous to forest land may disappear from the area, but other species of rodents more adapted to grasslands will quickly move in.

Historically, when you get this type of simplification of rodent species a few predominate and grow to high density, setting the stage for the emergence of rodent-borne diseases, including Bolivian hemorrhagic fever. A change in urban rodent fauna was a major factor in the emergence of the Black Death of the 1300s. The high population of infected rats (with their fleas) in the overcrowded, unsanitary conditions of Europe allowed the disease to flourish.

Changes in human ecology have contributed to the evolution of viruses and the spread of disease throughout history. Originally, bands of hunter-gatherers were small enough so that they supported only a few diseases transmitted among humans themselves. These viruses were mainly those such as herpes viruses which caused infections that were chronic or latent and thus could be maintained for a lifetime in the immune adults and passed on to susceptibles in the next generation. Of course, humans were infected with zoonotic viruses such as Rift Valley fever and VEE spread from mosquitoes or hantaviruses and arenaviruses spread from rodents. But several thousand years ago when humans began to congregate in larger groups, a new kind of virus began to emerge which was transmitted directly from person to person, such as measles virus. In the case of measles, as with other viruses such as influenza, the size of the potential host population is critical to survival. For a virus like measles to maintain itself, the population has to grow—children born or people moving in from places that haven't seen measles—to keep replenishing a supply of hosts susceptible to the disease. Once you've had measles, you're immune for the rest of your life and therefore useless to the virus.

This was discovered years ago when measles would sweep through an island population, rendering almost everyone alive at the time immune to the virus, thus causing the virus to die out. Then, years or decades later, when traveling sailors would reintroduce the virus to the island, all the people born since the last outbreak would get the disease, become immune, and the cycle would begin again. It can be shown that about half a million people are needed to provide the newly susceptible cannon fodder for continuous transmission of measles virus. The emergence of cities, with larger populations of hosts for viruses to work with and crowded conditions that make it easier for viruses to come in contact with more hosts, has been a tremendous boon to the virus world, rivaled only by our increased mobility. There is no reason why other viruses such as Ebola or the newly discovered paramyxovirus cannot adapt to human populations today. In fact, the catalytic event may well prove to be the megacity, which may provide viruses with the critical oppportunity, just as the emerging cities of the Middle East did for measles some 5,000 years ago.

There are other, more subtle "hot spots." In the United States, the spread of disease among young children in day-care centers has been well chronicled. Few species are as germ-friendly as young humans: they will touch and taste anything; they excrete waste in less than the most controlled, sanitary conditions; they have relatively immature immune systems. But the danger of communal living is not limited to children. As life expectancy continues to rise and the world's population ages, then nursing homes will grow more crowded and we will see similar problems on the other end of the life cycle.

These considerations lead inevitably to the question of how many humans can share the same turf. In 1798, the English economist Thomas Malthus published a landmark piece entitled, "An Essay on the Principle of Population," which made some rather dire predictions about overpopulation. Malthus' timing may have been off, but not his deductions. You can't tell me that there isn't a maximum population for the earth.

The area at the foot of Mount Elgon in western Kenya, just off the road toward Uganda, is one of the most densely populated areas on earth. You won't find any skyscrapers there, no major highways, but better nutrition and health care have allowed more people to survive. The subdivision of the agricultural land as it's inherited from generation to generation has resulted in smaller and smaller parcels supporting more and more people on them.

When that portion of the world, or India, or China, has reached its carrying capacity, what do we do? There is bound to be a growing rift between the haves and the have-nots, of which Dr. M.'s two medical practices in Bangkok are just a small example.

What does all this have to do with disease? The kind of population density we're seeing and will be seeing in the megacities of Cairo, São Paulo, Rio, Mexico City, Calcutta, Bombay, Jakarta, Beijing, is an invitation to epidemic disease. We're creating cities in areas where there's no money for surveillance or for effective communication, no medical infrastructure to recognize new diseases as they crop up or treat the old ones as they reappear. We're creating a recipe for disaster.

Put people close enough together in their slum or shantytown living conditions so that when one expires air, someone else inspires it, where they defecate in rivers where downstream others are getting their drinking water and washing their clothes, and that's going to be just like compressing uranium to make an atomic bomb. We're compressing people and germs together in a way that's bound to be explosive.

There are a host of politically charged issues that directly relate to our ability to curb disease transmission. On a practical level, if you travel in Africa today, for example, you run just a miniscule risk of contracting Ebola. But if you're in a car accident there—which is a very common occurrence, given the quality of roads, vehicles, and driving practices on the continent—and need a blood transfusion, it is not unlikely you will acquire HIV with the plasma.

And the problems do not lie just in the Third World either. In 1996, so-called mad cow disease broke out in England. Technically known as bovine spongiform encephalopathy (BSE), it is not a virus as far as we can tell but a prion, a microscopic protein similar to a virus but lacking DNA or RNA. Prions can induce changes in molecules which destroy neurons in the infected cow's brain. There is no blood test for it. The cattle disease resembles a similar one in sheep known as scrapie (so named because of the way infected sheep would scrape their hides until they were raw). Scrapie has been known since the 1940s, but BSE was unheard of until 1986. It may be related to kuru and Creutzfeldt-Jakob disease, two human degenerative diseases of the brain which are 100 percent fatal.

Although it is not yet proven, the generally accepted theory is that the agent that causes scrapie made the jump in species to cattle when infected sheep parts were added to cow feed without extensive processing. It has long been common practice to supplement farm animals' basic meal of grains, soy, corn, and the like with offal—leftovers from other animals that have been butchered. As Dr. Raymond L. Burns of the Kansas Department of Agriculture put it for *The New York Times*: "We use everything but the squeal, the cluck and the moo."

Will BSE prove able to jump species into humans? We're talking about a huge pucker factor here.

Part of the problem is our agricultural monoculture. Our huge stockyards and feeding pens, where hundreds or thousands of cattle are kept shoulder to shoulder, are the animal equivalent of our megacities. And the common practice of grinding up these leftover cattle and sheep parts and putting them into the feed of other domestic animals for economic reasons has a potential to recirculate and amplify the problem.

These decisions go back to evolutionary principles. By trying to maximize our yield, by continually crossbreeding cattle with the sole aim of coming up with the best meat producer in a short period of time, we're artificially limiting natural genetic variation and in the process creating a very special niche for microorganisms to flourish. If a naturally occuring or variant virus strikes a lonely cow in a field in Montana, the virus may infect a few other cattle and die off, but the mega-agribusiness provides a new niche with potentially susceptible animals in extremely crowded conditions, livestock that is trucked to even more crowded feed lots. Avian influenza can strike the similarly constituted modern commercial chicken houses and literally wipe out thousands of birds overnight. Finally, this crowded monoculture of animals provides a multiplier effect for human pathogens, such as toxigenic *Escheria coli* or salmonella in cattle or salmonella in chickens.

The same principles of genetic homogeneity apply to many of the specialized crops of the green revolution that we are counting on to prevent starvation in the years after 2000. A well-known historical example is the Irish potato famine of the nineteenth century. Potatoes were brought from South America, where they originated, to provide high-calorie food sources for the peasants of Europe. But limited varieties were introduced, setting up the crops for disaster later with importation of a fungus that attacked the few varieties grown in Ireland. At the time of the conquest of the Americas, there were 3,000 species of potato in the Andes and a proportion of these are genetically resistant to the particular fungus that caused the potato famine, thus providing a natural safeguard against catastrophic crop destruction.

Scientific sophistication, in and of itself, isn't the complete answer. I'll admit there may be economic advantages to incorporating anti-biotics in the feed of calves, but as a society we have to decide if we're willing to put up with some of the consequences, which very clearly involve an increase in the circulation of antibiotic-resistant bacteria.

We have already made that choice with tetracycline and many other antibiotics, in essence by not choosing. When tetracycline started being added to feed, no scientific analysis was carried through to a policy level. By default, then, the drug stayed in the food indus-tries and this, combined with promiscuous clinical prescribing, re-sulted in an increase in tetracycline resistance.

An additional problem is that this resistance can be transferred to other bacteria, so overuse of an antibiotic this way leads to an envi-ronmental increase in the pool of bacteria with the genetic ability to transmit resistance to their fellow bugs.

We have been depleting our armaments through other examples of overuse, or just poor choice of use. Not infrequently, physicians prescribe an inappropriate antibiotic, fail to obtain proper cultures, and at times even give antibacterial drugs like penicillin for viral infections, a useless excercise.

As a consequence of all these actions, some of our formerly potent antibiotics, including tetracycline and chloramphenicol, had lost their ability to fight the common strep and staph infections (like strep throat) by the time I was an intern in the late 1960s. Even worse, the process continues with newer drugs and has resulted in some com-mon and troublesome bacteria that defy treatment by any antibiotic. Even pneumococcus, one of the commonest childhood pathogens, has now become resistant to penicillin, unthinkable a few years ago.

Virtually the only up side to the story is that chloramphenicol was made available to the poorer nations of the world, where it would otherwise have been too expensive. It was a very costly drug that was widely overused in this country. In addition to development of bacterial resistance, one in 50,000 people who take it get a usually irreversible, often fatal aplastic anemia. This made it unmarketable in this country and prices plummeted. Still in patent, Parke-Davis began to market it extensively in the Third World. It was excellent against diseases prevalent in those areas—like typhoid, against which resistance didn't develop until the mid-1980s.

Contrary to medicine's own optimistic projections in the 1950s and 1960s, we're not going to end infectious disease any more than detectives and our other law enforcement brethren are going to stop violent crime. The best we're each going to do is to keep fighting with the best resources we can muster.

Viruses are all around us, and many of them are real threats. That doesn't mean we have to race to develop a vaccine for each one; we just don't have the money or manpower to even consider that. But I feel strongly that we do need to understand how they work, study them in detail, and at least get a grip on the principles of what to do about them. The first step in this effort is to work out the basics of a vaccine or drug therapy, not for each virus, but for selected members of each viral family. Then, when a serious and urgent need arises, we'll be that much farther along the learning curve.

Ribavirin, which I believe holds much promise, was first tested at USAMRIID in the early 1980s against virus grown in vitro in cell cultures. It showed great potential against arenaviruses such as Lassa and the South American hemorrhagic fevers. We've now got to test it more extensively in real-world situations where it might be needed and used.

Is this science for the benefit of all, or an example of scientific colonialism? These are the questions we wrestle with constantly. Presently, ribavirin costs about $1,000 a treatment, so even though Lassa fever is endemic in Africa, it's too expensive to use there on a mass scale. Yet that's where the studies are being done. Is it fair to use these people as our human guinea pigs, to tantalize them with a treatment they won't be able to afford? I don't claim there are easy answers.

But consider this: it makes sense to test the drug in places where the diseases it hopes to fight are most prevalent. And in a few years, hopefully around the time we know more about the drug, ribavirin will be less expensive, it will go out of patent, and it will be manufactured in mass quantities in places like Russia, China, and Brazil. And

I hope it will have a large impact on public health. In science, we always have to take the long view. Even in the short term, however, I think it is important that scientists, doctors, and technicians who are laying their lives on the line to study dangerous diseases should be afforded all the protection we can manage for them. And if a ribavirin-sensitive disease happens to erupt here at home, perhaps we'll be able to neutralize it before it spreads.

On the other side, though, I see dangerous trends in health care even in our own country that mitigate against surveillance and early intervention. Health Maintenance Organizations and "managed care" are largely creatures born of economic factors in society rather than the best medicine. Although money is already being saved, there are negative economic incentives to performing procedures or running laboratory tests that would enable the physician to make a definitive diagnosis. I can't believe it is better for the patient, or even cost-effective, to not make a specific etiologic diagnosis, whether in an epidemic setting or with an individual patient.

We know, for example, that lymphocytic choriomeningitis virus in a pregnant woman can cause hydrocephalus in her child. We don't know much about the prevalence of the virus in this country and therefore know little about what other implications it may have for public health.

I've talked to doctors at infectious disease meetings about running diagnostics for this virus. But their response is: "If I do that, it pushes up my cost profile with the HMOs that refer patients to me. If I get a high cost profile, then I'm going off their list." In this case, there may not be treatment implications for that specific patient, but there may be enormous public health implications and valuable information in learning about emerging diseases as well.

If you look at the U.S. mortality tables from 1900 and compare them with the mortality tables today, you won't find many similarities. The big killers at the turn of the century were mainly infectious diseases. Now they're chronic ones like heart disease and cancer and diabetes. And just as we've made impressive progress in the diagnosis and treatment of the old killers, we'll eventually make similar progress in cancer, heart attacks, and stroke.

But as the old expression goes, no matter what we cure, there will always be a leading cause of death, and infectious disease could well return to the top of the charts if we don't start putting serious resources into their prevention.

My thirty-year career as a virus hunter has taught me that there are many tests and trials along the way, as well as dangerous and deadly curves when we least expect them. I also know there is great reason for hope too. We're not going to cure death, but the effort we put into fighting it off, and the reverence we place on the value of every human life, says more about us as a species than anything else I can think of.

I think it's essential that we be prepared. There's been a lot of discussion among experts and commentators, book and magazine journalists about how much of a threat these so-called emerging viruses represent. The current vogue is that they're not as much of a large-scale, Andromeda Strain-type threat as some people have suggested. And they could be right. Maybe nothing's going to happen. But there is something terrifying about the fact that nothing can stop the implacable evolution of these viruses as they test, through mindless mutation, ever more strategies to facilitate their survival, a survival that just may represent disease and death for us humans.

Maybe no deadly pandemic will occur. But I wouldn't want to bet my life on it.

Formerly Chief of Special Pathogens at the Centers for Disease Control (CDC) in Atlanta, Georgia, and Chief of the Disease Assessment Division at USAMRIID, C. J. PETERS has worked in the field of infectious diseases for three decades with the CDC, the U.S. Army, and the U.S. Public Health Service. He was the head of the unit that contained the outbreak of Ebola at Reston, Virginia. He is currently Director, Center for Biodefense, University of Texas Medical Branch, Galveston, Texas. MARK OLSHAKER is an award-winning filmmaker and writer who coauthored the bestselling nonfiction book *MindHunter*. His acclaimed novels of suspense include *Einstein's Brain, Unnatural Causes, Blood Race*, and *The Edge*. He is married to Carolyn Olshaker, an attorney, and lives in the Washington, D.C., area.